Plastics Extrusion Technology Handbook

Plastics Extrusion Technology Handbook

by

Sidney Levy, P. E.

INDUSTRIAL PRESS INC.
200 Madison Avenue
New York, New York 10157

Library of Congress Cataloging in Publication Data

Levy, Sidney, 1923–
    Plastics extrusion technology handbook.

    Includes index.
    1. Plastics—Extrusion—Handbooks, manuals,
    etc.   I. Title.
TP1175.E9L48   668.4'13   81-6587
ISBN 0-8311-1095-3   AACR2

PLASTICS EXTRUSION TECHNOLOGY HANDBOOK—**FIRST PRINTING**

PREFACE

The art of plastics extrusion has developed over the last century from one using crude presses and questionable materials to a highly sophisticated process for manufacturing a wide variety of important products. The advances in the art range from better materials to an understanding of the flow characteristics of high polymers. Better machinery and control systems have made possible closer control of the process and the manufacture of product with precise dimensions and good material characteristics.

The subject has been covered extensively in the literature from the characteristics of screws to the rheology of materials. This text draws on the extensive literature for background. It is directed to the technology of the process—the machines, dies, and auxiliaries used to manufacture product. It provides an introduction to the mechanisms whereby extruders work and how plastics flow through dies. Its primary purpose is to provide the information needed to design, install, and operate plastics extrusion systems to make product. While extruders are used in other plastics processes such as injection molding, blow molding, and compounding, these subjects are not covered since the extruder acts only as a melt-generating unit in those processes.

It is hoped that this presentation on the technology of plastics extrusion will provide a useful guide to those working in the field. The extensive literature references will enable those with a need for more detailed information to go to other sources to supplement this handbook.

INTRODUCTION

Plastics extrusion is one of the most important processes used in the plastics industry. Fully 60% of all plastics material passes through an extruder on its way to conversion to a product. Some of this processing is to pelletize, compound, and color raw materials, but the bulk of the processing is for conversion to finished product or semifinished stock shapes. An examination of some classes of products made by extrusion will give an idea of the scope of the process.

Products made by extrusion range from simple shapes such as sheet and film to rod, pipe, and tubing and complex profiles, which are used for both industrial and commercial applications. Gaskets, house siding, structural furniture parts, decorative moldings, and window and door tracks are some examples of widely used profile extrusions. A variation of straight extrusion called crosshead covering is used to make insulated wire, decorative foil moldings, and plastics-covered towel bars. The coextrusion of several different materials to make laminated sheet and dual durometer parts is another extension of the process capability. The text will discuss these and many other applications for the plastics extrusion process.

The wide range of products results from a fundamentally simple technique that consists of heating a plastics material to melt it, forcing the melted material through a shaping die, and subsequently cooling it while holding the shape. In some instances a post die-forming process is included, which extends the scope of the product capabilities. From rather simple and crude beginnings as an adaptation of the rubber and clay products extrusion equipment, the process has become more sophisticated and complex. This resulted from the need to process new and more difficult materials, and the need for better process control. The improved process control is necessary for precision in dimensional control and control of the physical properties of the extrusions.

The text will cover equipment and process as well as the theory of operation of both machines and tooling. The process control techniques for shape control and uniformity essential to successful operation are discussed in detail. The productivity of specific equipment used on a wide variety of products is described so that appropriate plant equipment and tooling can be selected to make a product at the desired rates. Plant design and operation is described to provide the plant designer or the plant operator with sufficient information to understand the manufacturing methods and the limitations of the extrusion process.

Important subjects to be discussed are (1) design of screws and barrels for the extrusion machine; (2) heating and cooling systems for the extruder, the dies, and the cooling equipment; (3) shape-holding equipment such as cooled rollers, vacuum sizers, and similar devices; (4) machine drives; (5) pullers; and (6) special equipment for specific products including dies. Each of these has a bearing on product quality, productivity, and costs. The diversity of ways in which each of these operations can be done lends great flexibility to the plant design but requires studied decisions as to the most appropriate devices to be used.

In a general text such as this, it is impossible to cover all of the relevant extrusion process technology. The intent is to present material which will lead to a reasonable understanding of the process. This will be amplified by examples of specific production lines used to make a number of widely used products. The philosophy of design involved will give guidance to the plant designer and operator when faced with extensions of the techniques to new materials and new products. The operations will be described with some of the typical problems associated with them to indicate how such operational situations can be analyzed and corrected.

Materials will be discussed with respect to the way in which they process and what properties are required for successful extrusion of certain products. Not all plastics materials can be used in making all extruded shapes. The selection of materials and their interaction with product specifications and processing requirements is a major activity involved in plant design and operation.

This volume should be a useful guide to tooling and manufacturing engineers, equipment designers, plant operators, and product designers who need knowledge of the plastics extrusion process in their work. The material covered is broad enough to supply each of these interests with the basics of the process and the references cited will enable the reader to extend his or her knowledge in specific areas. Extrusion personnel can use the book to understand, design, and operate the extrusion systems for efficient production and to upgrade the plant performance by the use of the best available technology.

CONTENTS

Plastics Extrusion Technology Handbook

Chapter One
Fundamentals of the Extrusion Process

Plastics extrusion is a process for producing a plasticized mass of polymer compound, forcing it under pressure through a shaping die orifice, and subsequently setting the shape by means of a cooling and shaping system. The shaping process in the die, which is analogous to squeezing toothpaste from a tube, is the key part of the process and will be discussed first.

Figure 1-1 shows an idealized extruder orifice with the material passing through. It is assumed that the material has been completely plasticized or melted and that the die orifice is at a temperature that will not cause cooling of the material. What is of interest is how much material will flow through the orifice in response to an applied pressure and how uniform the flow will be in various parts of the orifice. The first factor will determine the throughput and the second will determine the effectiveness of the orifice as a shaping device. Both of these factors are controlled by the material rheology.

The rheology of a material is the manner in which it deforms in response to an applied stress. In the case of plastics in the melt condition, the stress applied is pressure, and the response is continuous deformation and flow. The flow is dependent on the shear characteristics of the material, and this response can be of several types, as illustrated in Fig. 1-2. If the material has a constant viscosity, i.e., constant ratio between shear stress and shear rate, it is called a Newtonian fluid. The shear stress and shear rate can be defined in conjunction with the illustration in Fig. 1-3 which shows a material being sheared between two parallel plates:

$$\text{shear stress} = \frac{\text{shear force}}{\text{shear area}} = \frac{F}{A}$$

$$\text{shear rate} = \frac{dv}{dh} = \frac{V}{h} \text{ (for a Newtonian fluid)}$$

$$\text{Newtonian viscosity } \mu_N = \frac{\text{shear stress}}{\text{shear rate}} = \frac{F/A}{V/h}$$

$$\text{general viscosity } \mu = \frac{F/A}{dv/dh}$$

Most plastics are non-Newtonian in their characteristics. This is to be expected from the fact that they are made up of long-chain high-polymer materials. The viscosity of the materials will change with the amount and

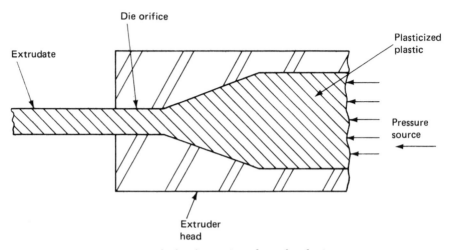

Fig. 1-1. Idealized extrusion scheme for plastics.

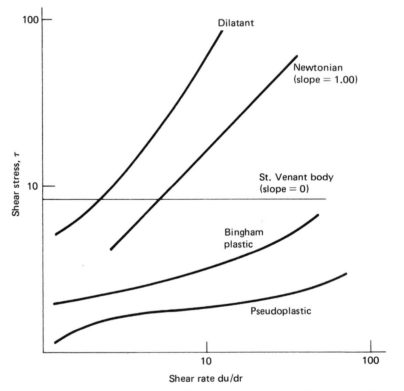

Reprinted from E. C. Bernhardt, Processing of Thermoplastic Materials,
Reinhold, New York, 1959, p. 17.
Fig. 1-2. Flow curves for materials with different rheological characteristics.

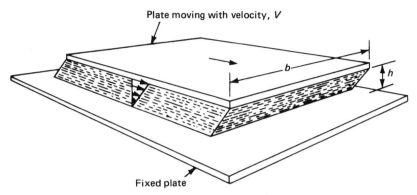

Fig. 1-3. Diagram for simple shear for a Newtonian fluid between parallel plates.

duration of the shearing forces, which have a tendency to align the polymer molecules in the flow direction and hence reduce the resistance to flow. As a result, the flow of the polymers through the die orifices is quite complex and requires analysis to determine the true shaping effects.

Data on the response of plastics materials to flow are generally reported in the literature in the form shown in Figs. 1-4 and 1-5. These are, respectively, the shear rate/shear stress curve and the shear rate/

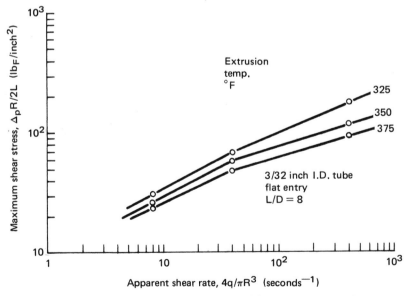

Reprinted from E. C. Bernhardt, Processing of Thermoplastic Materials, Reinhold, New York, 1959, p. 656.

Fig. 1-4. Plot of relationship between shear rate and shear stress.

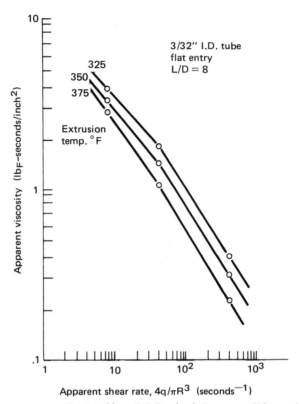

Reprinted from E. C. Bernhardt, Processing of Thermoplastic Materials,
Reinhold, New York, 1959, p. 657.
Fig. 1-5. Plot of apparent viscosity versus shear rate.

viscosity curve for polyvinyl chloride, a material which is widely used. An examination of the curves reveals that, indeed, the viscosity is lower for high shear rates and the shear stress is lower for higher shear rates. The effect is to modify the flow through the die orifice.

The flow modification caused by shear-dependent viscosity is shown clearly in Fig. 1-6. A Newtonian fluid would have a velocity profile flowing through the die as shown in Fig. 1-6a since the viscosity is independent of shear rate. With a material that has a viscosity dependent on the shear rate, this may be the initial shape of the velocity profile, but it will be modified rapidly since the material in the vicinity of the walls is sheared at a much higher rate than the material in the center of the extrudate stream. The ultimate velocity profile may be the one shown in Figs. 1-6 b, c, or d, or in the limit e. This results from higher dependence of viscosity on shear rate and the development of lower viscosity flows

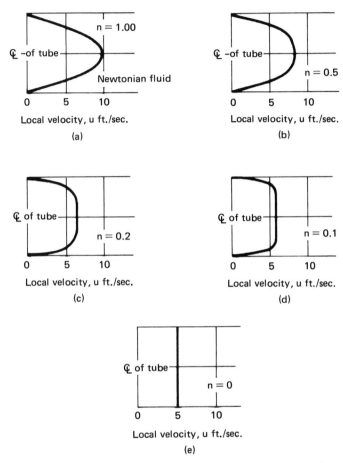

Reprinted from E. C. Bernhardt, Processing of Thermoplastic Materials,
Reinhold, New York, 1959, p. 51.
Fig. 1-6. Velocity profiles of power-law fluids flowing inside round tubes in laminar flow.

near the orifice walls. The more dependent the viscosity is on the shear rate, the more closely the velocity profile will conform to Fig. 1-6e, a condition described as plug flow.

Those polymers that exhibit this property to a high degree are most readily extruded to the shape of the die orifice. The property is called thixotropy, and the polyvinyl chloride resin (whose rheology curves are shown in Figs. 1-5 and 1-6) is one that is easy to shape. In addition to the shear, the effect of temperature on the viscosity must be taken into account in order to understand flow through the orifices. Die swell, which is a result of the viscoelastic nature of the polymer melt, is another

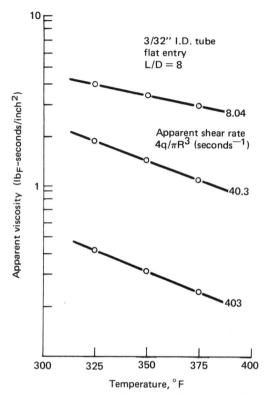

Reprinted from E. C. Bernhardt, Processing of Thermoplastic Materials,
Reinhold, New York, 1959, p. 657.
Fig. 1-7. Effect of temperature on apparent viscosity.

variable. Energy, stored as pressure in the material going through the die orifice, is released on exiting and exhibited as an increase in the cross-sectional area of the extrudate. These effects are shown graphically in Figs. 1-7 and 1-8.

The reduced viscosity resulting from elevated temperature increases the average flow rate at constant pressure. In some cases the control of melt temperature can be used to balance and control the output from various areas of the die orifice by changing the local temperatures. One complication results from the temperature–viscosity dependence because the shearing effects are heat generating. Where there is a high shear rate, the material will heat up and reduce the viscosity. This adds to the viscosity reduction caused by the shear alignment of the polymers and will accentuate the transformation from parabolic to plug flow. It can also cause localized degradation of heat-sensitive polymers.

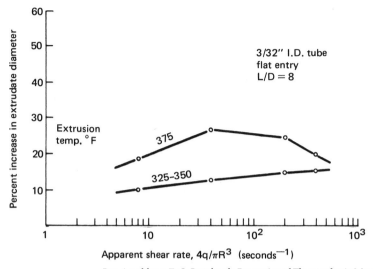

Reprinted from E. C. Bernhardt, Processing of Thermoplastic Materials,
Reinhold, New York, 1959, p. 657.
Fig. 1-8. Effect of shear rate on die swell for two temperatures.

The die swell shown in Fig. 1-9 is a consequence of two mechanisms, both related to the properties of the polymer melt. Because of the viscoelastic nature of the polymer materials, pressure energy is stored as omnidirectional strain in the melt. This is converted to stresses perpendicular to the flow direction in the orifice. When the resin exits from the orifice, the stresses produce the strain exhibited as die swell. The other mechanism involved in die swell is a frictional one. This is exhibited by polymers which have a high coefficient of friction against the material of which the die orifice is made. By restricting flow of the boundary layer of the melt stream, the effect is to restrain flow at the die/material interface

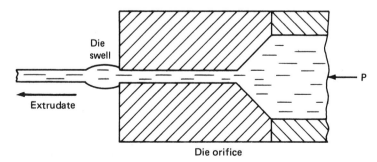

Fig. 1-9. Schematic of die swell in the flow of plastics.

while the bulk of the stream flows more freely. As a result, the extrudate will swell or bloom. This effect is more evident with materials like polyurethane which have high friction coefficients. It is almost nonexistent with materials such as polyethylene which are low-friction-coefficient materials against most die materials. The magnitude of the die swell can exceed 100% of the die orifice dimensions. This can be reduced by changes in material temperature, lubricant levels in the polymer melt, die land-length changes, and changes in the entry angle into the die orifice. Changes in the surface coatings on the die orifice change the swell ratio. It will also change the rate for the onset of melt fracture, a related phenomenon.

The phenomenon of melt fracture or melt instability is one in which the material coming out of a die orifice becomes so irregular or rough that the product is unusable. Melt fracture material is shown in Fig. 1-10, and the curve in Fig. 1-11 indicates that at some shear rates the effect will occur. It usually occurs at a high swell ratio, and it is a specific characteristic of a particular plastics formulation and die configuration. The initiation of melt fracture will also depend on such factors as temperature, lubricant levels, land length, entry angle, and surface quality in the die orifice. The appearance of melt fracture is the

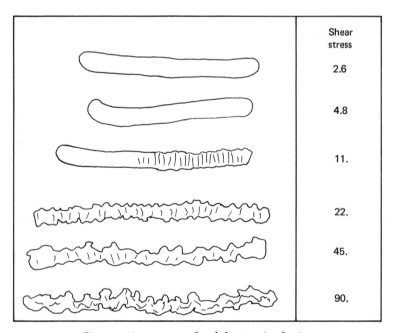

Fig. 1-10. Appearance of melt fracture in plastics.

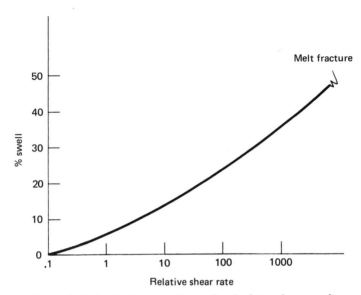

Fig. 1-11. General relationship between die swell and relative shear rate, linear plot.

determining factor in the maximum possible flow rate through a die. This, in turn, determines the necessary pumping action for the extrusion process. (For additional information see Ref. 1.)

The plasticating and pressure-generating functions are the next part of the process to be examined, since there is a high degree of interaction between the die orifice and the pressure-generating unit in the process. There are many methods for performing both functions. Figure 1-12 illustrates a plunger and heated cylinder which is used to extrude. This method is used in melt rheometers to check the properties of materials and has been used in some production equipment. To get continuous streams of material from what is a batch process, it is necessary to equip the unit with several plungers and a valve arrangement so that the cylinders can be recharged and discharged in sequence, as shown in Fig. 1-12b. Most extrusion pumps require a continuous output so that valves are necessary in the plunger extruder.

Figures 1-13 a, b, c, and d illustrate several types of pumps that are used for extrusion. Figure 1-13a is a screw pump of the type that is incorporated into single-screw extrusion machines. Figure 1-13b illustrates a twin-screw pump used in a twin-screw extruder. It has the desirable feature that the delivery rate is relatively independent of the output pressure. Gear pumps, such as shown in Fig. 1-13c, are used for low-melt-viscosity materials and in conjunction with single-screw pumps

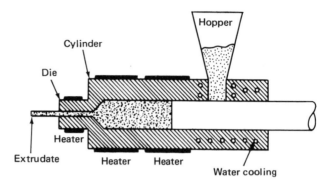

Fig. 1-12a. Schematic diagram of simple ram extruder.

to control output rates. A different approach to pressure generation is shown with the Maxwell melt elasticity pump (Fig. 1-13d), which is an example of one unusual polymer pump depending for its operation on the special characteristics of polymer melts.

The extruder performs three actions in the process. It melts or plasticates the material, it generates pressure on the material to force it through the die orifice, and it shears and mixes the material. Each of the systems illustrated in Figs. 1-13 a, b, c, and d does these in a different manner and in doing so imparts different characteristics to the melt. The ram extruder heats primarily by conduction of heat from the walls of the extruder and by the use of heaters that preheat the material before it is placed in the cylinders; it generates pressure by direct hydrostatic effects.

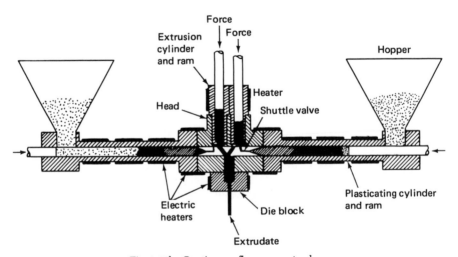

Fig. 1-12b. Continuous flow ram extruder.

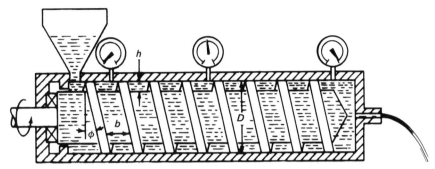

Fig. 1-13a. Pressure development by shear of a viscous fluid between screw and cylinder.

The mixing and shear are limited to that occurring as a result of the direct flow and entry into the die orifice. The other extrusion machines operate in much more complicated modes.

The single-screw pump is dependent for its pressure generation effects on a mechanism called drag flow, which will be covered at length in Chapter 2. The pressure buildup is the result of interaction between the material and the screw and barrel wall, and the delivery rate is a complex function of material temperature, barrel temperature, and shear rate. The melting of the polymer is done by a combination of heat passed through the barrel wall and the effects of shear heating of the material. There is extensive shear on the resin, and extensive mixing as the material passes through the machine. The extent of the shear, the material temperature, and the mixing are highly interdependent and cannot be separately controlled. As a result, it is sometimes difficult to get stable operation at high throughput rates.

Twin- or other multiple-screw pumps use intermeshing corotating or counterrotating screws to build up pressure on the polymer melt. While the clearances between the screws are not close, the machine is a

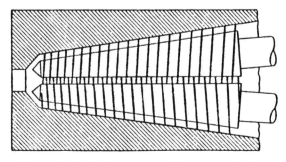

Fig. 1-13b. Conical converging twin screw extruder.

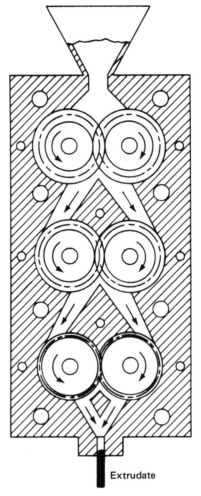

Extrudate

Fig. 1-13c. Diagram of a gear pump extruder.

positive pumping device that does not depend on material characteristics for the delivery rate. It is not as positive as the ram extruder, but it is very stable in output, independent of the effects of shear on the resin passing through the pump. Since the functions of pressure buildup and shear mixing are essentially independent, it is possible to independently vary each by appropriate screw design and machine control. The heat to melt the material is provided through the barrel wall with a relatively small amount generated by shear heating. The material is actually conveyed in a moving cavity formed by the screws and the barrel wall, with a small

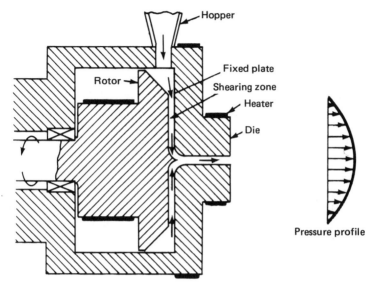

Fig. 1-13d. Diagram of an elastic melt extruder (Maxwell).

amount of shear at nip formed between the screws. The twin-screw pump is used with materials that exhibit high shear sensitivity because the shear and delivery can be independently controlled.

The gear pump is a positive displacement device which is used as an extrusion pump. Since it uses intermeshing members it can develop high delivery pressures at constant delivery rates. Because the internal volume of the pump is small for the effective delivery rates, it is difficult to heat the polymer through the pump walls. As a consequence, gear pumps are usually melt-feed pumps, and are more frequently used to control the output rates from other melting devices such as single-screw pumps. Two other characteristics of the gear pump limit its usefulness. One is the fact that it cavitates readily on the inlet with high viscosity melts; the other is that the intermeshing teeth impart locally high-shear rates which create problems with shear-sensitive resins. They are generally used for relatively low-molecular-weight, low-viscosity resins, such as those used to make textile fibers or adhesives.

The elastic melt pumps use the normal component of force developed as a result of the polymer shear effects on viscoelastic materials. The units are decidedly more shear sensitive than any of the other pumps discussed. In order to get any pumping action, the material must be sheared at a high rate. The melt elasticity pumps do not generate very high pressures. In order to use the units for production, they are equipped with pressure-generating pumps, such as single-screw pumps or gear pumps, after the elastic melt unit. The very high degree of shear, which is uniformly imparted to the entire melt, makes very uniform melt quality. The elastic melt pumps are used where mixing and color dispersion are necessary or in other cases where a high degree of control over melt quality is essential to the process. They should not be used for shear-sensitive materials, which can be degraded by the action of the pump.

As will be discussed in detail in the chapter on machine design and operation, actual production equipment requires the use of combinations of different mixing and pumping actions to achieve proper melt conditions for successful extrusion operation. The interaction of the die orifice with the extruder, particularly the single-screw machines, is important to the proper functioning of the machines. Matching die characteristics with the extruder function is one of the most critical elements in successful operation of a system.

The third stage of the extrusion process is the removal of the shaped extrudate from the die orifice, and the subsequent cooling while the part is maintained in the desired shape. A pulling device is required to move the extrudate away from the die orifice as it exits from the machine. The devices used for this purpose should be capable of adjustable but controllable pulling rates since the cross-section of the extrudate will depend on the relative rate of pull versus delivery from the die orifice. In some cases the pulling device can also provide some of the cooling necessary to set the extrudate.

A roll stack is used when sheet material is made from plastics by

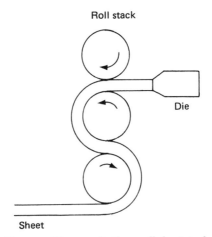

Fig. 1-14. Diagram of a three roll sheet stack.

extrusion. This is illustrated in Fig. 1-14. The roll stack performs several functions. It pulls the extruded sheet away from the die orifice at a controlled rate of speed which is matched to the extruder output. It also provides cooling of the sheet to set the extrudate. The wide ribbon of material coming from the die lips is wrapped around the rolls and pulled by them. In this case the roll stack also provides a dimension-setting mechanism for the sheet thickness by the setting of the distance between the roll nips. Most extrusion systems do not combine all of these functions in a single piece of equipment.

The pipe-extrusion operation shown in the diagram in Fig. 1-15 is more typical. The actual product pulling is done by a double-tracked conveyor unit with an adjustable speed drive. The cooling is done by the use of water tanks. In some operations a vacuum sizing tank is used to make an initial set on the outside dimensions of the pipe before it enters the remainder of the cooling tanks.

The equipment used to shape, hold, and cool the extrudate after it emerges from the die orifice constitutes the major design effort in a plastics extrusion line. In many cases the post die tooling has much more effect on the shape and quality of the extrusion than the extrusion die. The nature of the equipment used depends to a substantial extent on the processing properties of the material and the type of product made.

The previous discussion on materials must be amplified to show what processing characteristics are required for good post die operation. Simple shapes such as sheet and strip do not require special melt properties such as high-melt strength, since the resin flowing out of the die can be dropped directly onto the first roll of the roll stack where it is cooled sufficiently to be handled in the remainder of the line. Higher melt strength (resistance to easy flow deformation at the melt temperature) is required for most other operations such as pipe and profile extrusion. The extrudate should be capable of supporting itself over the distance between the die and the first cooling section of the line. In thin sections, such as tubing made with a sizer, retention is not as essential.

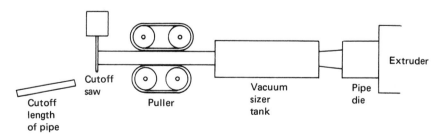

Fig. 1-15. Schematic of a plastics pipe extrusion line.

When special profiles are extruded, it is highly desirable that the melt exhibit a high degree of thixotropy in order for the extrudate to hold the shape imparted by the die until it can be supported in the cooling jigs and fixtures in the extrusion line. Materials which do not change viscosity by several orders of magnitude as soon as the shear forces are removed are difficult to extrude into complex profiles. Some materials, such as polyvinyl chloride, exhibit this property to a high degree, while other materials such as the polyolefins and polyamides do not. The former have been extruded into very complex shapes, while the latter are rarely shaped by the extrusion process.

Another characteristic of the polymers substantially affects the extrusion system performance in the post die operations. This is the degree of crystallinity and the rate of crystallization of crystalline polymer. When the polymers cool they will crystallize and the change of morphology will cause sharp changes in the contraction rate of the extrudate. If cooling is not applied symmetrically to the extrudate, the shape will be severely distorted. In addition, the extrusion may warp, twist, and generate a bow or sweep condition that will make the product curved and crooked instead of straight and uniform. Highly crystalline materials such as acetals, polyamides, and some polyolefins are very difficult to handle in the cooling and shaping operation.

The thermal stability of the polymers is a factor in extrusion which is sometimes overlooked. It is not only important that the material be sufficiently stable to pass through the extruder without any significant loss in properties; it is also necessary that it be capable of withstanding the heating effects of high shear stress in some dies. Regions of high shear in some dies can raise the temperature of polymers with high frictional heat characteristics well above the incoming melt temperature. The effect of the high heat can be either degradation and burning, as exemplified by rigid PVC, or it can result in chain scission and the formation of low-molecular-weight products, which are typified by methyl methacrylate plastics. While the effects of the two degradation mechanisms on die performance are quite different, each is deleterious to die output.

Table 1-1 is a summary of the handling characteristics of a number of commonly used materials which are important in extruding plastics shapes. Not all of the considerations are important for sheet and film and other products. It does indicate, however, what the material limitations can be for use in extrusion.

In many extrusion lines other operations are performed on the extrudate on line to make more complete product. These operations include forming, punching, and embossing as well as a variety of decorative

Table 1-1. Low-Melt-Index Materials and Characteristics Affecting Profile Extrusion[a]

Material	Thixotropy	Melt Viscosity	Temperature Setting Range	Frictional Heat	Melt Elasticity
Rigid PVC	1	1	3	4	4
Flexible PVC	2	2	3	2	3
Acrylic	3	2	3	3	6
Modified PPO	2	3	2	2	5
ABS	3	3	3	3	4
Impact styrene	4	4	3	2	4
Polycarbonate	4	3	4	4	4
Thermoplastic polyester	4	4	2	3	4
Polypropylene	6	4	3	2	4
High-density polypropylene	6	5	5	2	2
Polyethylene	7	5	6	2	4
Polyamides	8–9	7	3	2	3

[a]Numbers represent the relative difficulty involved in extruding profiles of the materials. A low number, on a scale of 1 to 10, indicates that particular characteristic of the material makes it relatively easy to form.

operations. These are not the primary operations of the process, but it is very common to have one or more of these steps incorporated into a line. Here, again, there must be some consideration of the requisite properties in order to do the operation successfully. In the case of the embossing, it is necessary to know how much of a temperature drop is required to set a pattern in the extrudate. This will determine the heat removal requirements of the embossing unit. For forming operations it is important to know the forming temperatures for the material in order to design suitable temperature-controlling units to drop the extrudate temperature from the die exit melt temperature to the forming temperature. Punching also requires knowledge of the properties of the material. Many plastics are brittle when they are at room temperature and cannot be punched without cracking. It is important to note that the colder the material, the higher the load on the punch. Data on the punching characteristics of the material are important in order to be able to decide if the material can be punched on line and at what temperature.

Summary

This chapter has covered the basic principles of extrusion and has described the three key elements in the extrusion-forming process: forcing material through the extrusion-die orifice, plasticating the material, and post die handling and cooling. The requirements of each of these steps was discussed, and the effect of materials selection on each element pointed out. Each of these process parts will be discussed in detail in subsequent chapters to bring together the necessary design concepts for successful extrusion systems.

References

1. C. D. Han and R. R. Lamonte, "A Study of Polymer Melt Flow Instabilities in Extrusion," *Polymer Engineering & Science* **11**(5), Sept. (1971).
2. S. Levy, "Melt Rheology: Its Effect on Profile Extrusion Dies," *Plastics Machinery & Equipment* **7**(9), Sept. (1978).

Chapter 2
Extruder Design, Construction, and Operation

The extruder is the melt-supply unit in the extrusion process, and the design and operation of the extruder is a key element in a successful system. The function of the machine is to generate a supply of plasticated material with uniform temperature and composition at a constant and controllable rate. The machines have evolved from the crude devices that were converted rubber extruders to the sophisticated melt-generating equipment now in use. The construction of the machines has changed substantially from the original simple screw pumps.

The majority of the extruders in current use are single-screw machines. A great deal of investigation has been done on the theory of operation for these machines. Since the theoretical work has been substantiated by extensive experimentation, the operation of this device is well understood. Twin-screw and other multiple-screw machines are a more recent development. They, too, have been studied in order to understand the theory of operation, and, while the extent of the study is not as great as that of the single-screw machine, the operation of the equipment is well characterized. Other special extrusion devices such as the elastic melt extruder and gear pumps have been studied and characterized for the special applications for which they are suited.

The machines will be examined in regard to the general construction and mode of operation, and then the specific construction and design details of production equipment will be discussed. The description is intended to supply sufficient information to select a machine for a specific application and to properly understand the operation of the machine for the intended use.

Single-Screw Extruders

The appearance of a typical single-screw extruder is shown in Fig. 2-1 with the important parts labeled. The internal construction of an extruder is shown in Fig. 2-2 which shows the internal arrangement of the parts of the machine. The screw and barrel are the units that coact to convey, melt, and generate pressure in the plastics material. The drive system rotates the screw at controlled speed. The barrel is equipped with heating and cooling elements connected to temperature controllers. The functioning of the screw and barrel to pump the melt is dependent on the construction of the screw and on characteristics of the polymer material being extruded.

The screw-pump extruder operates on the principle of drag flow. Extruders have three different operating zones: the feed section, the

Fig. 2-1. Typical single-screw extruder of current design.

melting and plasticating section, and the metering section. Each of these zones has a different screw configuration in order to perform its role in the plasticating process. The feed section conveys plastics granules or powder from the feed throat of the machine to the region where the material begins to melt. Since the bulk density of the unmelted material is lower than the density of the molten polymer, the flutes in the feed

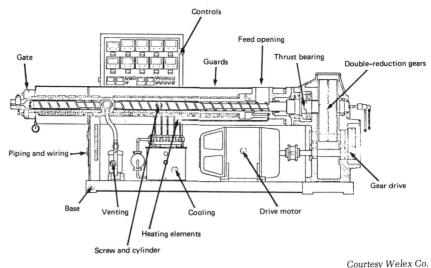

Fig. 2-2. Diagram of typical single-screw extruder showing construction details.

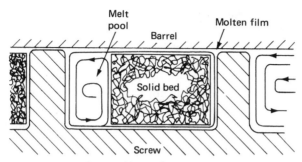

Reprinted from Polymer Engineering & Science **16**, 284, April (1976).
Fig. 2-3. Schematic of melting mechanism in a single-screw extruder (Maddock).

section are deep as compared with the flutes in the metering section of the screw. The transition section between the feed section of the screw and the metering section is the region where the plastics granules are softened by contact with the heated barrel of the machine and compressed to improve the melting of the resin. As the material starts to melt, a pool of material is formed in the rear portion of the screw flights. This pool of molten material is further heated by the shearing action on the resin caused by the relative motion of the screw and barrel. The melt pool increases in size until all of the material is melted. This sequence is shown in Figs. 2-3 and 2-4. Under proper operating conditions the melt pool will completely fill the screw flutes at the point of entry into the

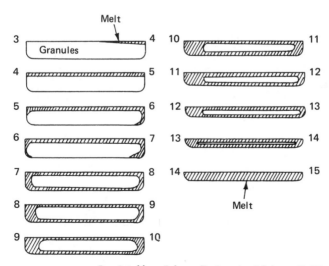

Reprinted from Polymer Engineering & Science **16**, 285, April (1976).
Fig. 2-4. Melting stages in the screw of a single-screw extruder.

metering section of the screw. The metering section of the screw then pumps the melt out at a controlled rate determined by the screw speed.

The pumping action of the screw is a drag flow phenomenon. The feed section represents a rather complex solids-conveying situation which depends on the coefficient of friction between the resin and the heated barrel. In most cases the forward feeding action is sufficiently positive to fill the screw flutes leading into the transition section without overfeeding. The melting in the transition section is controlled by proper adjustments of the barrel temperature so that completely melted polymer is delivered to the metering section. The final pressure buildup is done in the metering section by the drag flow mechanism operating on the fully plasticated melt.

The drag flow pumping mechanism depends on polymer wetting both the barrel and the screw in the melt pumping section. The pumping action is shown in the illustration in Fig. 2-5. In this diagram the screw is shown with the directions labeled. To clarify the analysis the screw is shown "unrolled" to provide planar geometry for calculations. It is apparent that movement of the "barrel" surface in the direction l will cause two components of force to be generated in the x and z directions on the material in the channel. This will produce a motion toward the exit section of the machine. In essence, the material is being pushed or dragged toward the front of the machine by the action of the angled

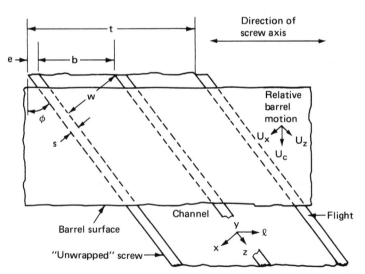

Reprinted from E. C. Bernhardt, Processing of Thermoplastic Materials, *Reinhold, New York, 1959, p. 167.*

Fig. 2-5. Flat plate model of a double-flighted single-screw extruder.

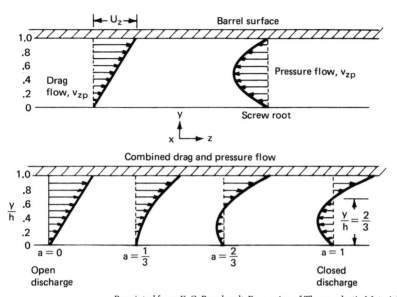

Reprinted from E. C. Bernhardt, Processing of Thermoplastic Materials,
Reinhold, New York, 1959, p. 179.
Fig. 2-6. Flow profiles down channel in a single-screw extruder.

(helical) flights. There are two directions of motion generated. One direction is across the screw flight and the other is along the flight. The component along the flight generates the drag flow along the screw.

As the drag flow generates pressure in the material there is a tendency for flow back through the open screw channel. This flow runs counter to the drag flow component and reduces the net movement of the material along the extruder. Figure 2-6 shows the material velocity profiles generated individually by the drag flow and pressure flow and then the combined flow under different discharge conditions: open discharge flow with no die restriction, flow reduced by one-third from the open discharge condition, flow reduced by two-thirds from the open discharge condition, and flow completely blocked. Under the open discharge condition there is no restriction at the end of the extruder barrel and no back pressure is generated. As a consequence, the flow is simple drag flow. As the material flow is restricted, back pressure is developed, and the flow profile includes the backward-tending component of the pressure-induced flow. When the flow is completely blocked, the integrated velocity profile for pressure flow is equal to the integrated velocity profile for drag flow for a balanced condition with no net output from the extruder.

Figure 2-7 shows the condition in a plane transverse to the channel,

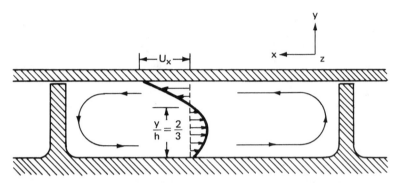

Reprinted from E. C. Bernhardt, Processing of Thermoplastic Materials, Reinhold, New York, 1959, p. 178.

Fig. 2-7. Flow profile in a transverse plane of the flight.

along with the streamlines that show the flow direction for the material during totally blocked flow. During partially blocked flow conditions the flow pattern will be tilted in a direction toward the exit section of the extruder, and the material will spiral through the flutes to develop a net output flow.

The melting of the polymer in the screw channels is a complex process. The initial contents of the channels are compacted granules of material which make intimate contact with the heated barrel wall. The conducted heat starts to melt the material. This results in a compacted

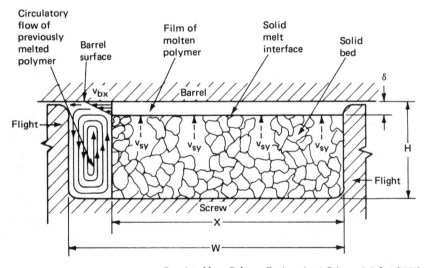

*Reprinted from Polymer Engineering & Science **9**, 2, Jan. (1969).*

Fig. 2-8. Idealized channel cross section showing melting mechanism.

mass of polymer against the rear part of the screw flight with a melted pool in front. This is shown in Fig. 2-8 where the melt pool and the solids bed are shown in the condition that would exist approximately halfway down the melting zone of the extruder. The melt pool is undergoing extensive shear mixing as a result of the drag melt action, and this shear action produces viscous heating of the polymer. In many polymers the frictional heat is the major source of heat needed to melt the material. In all cases, the shear mixing provides a significant part of the energy to melt the polymer.

Under proper operating conditions, the melting is complete at the end of the melting or transition zone of the extruder. The plastic that enters the metering zone of the screw will be completely plasticated. The metering zone is a melt pump which controllably delivers melt at constant rate to the die on the machine. If the machine settings are incorrect, or if the screw is not matched to the melting characteristics of the polymer used, there are some undesirable consequences. One of the common problems is poor melt quality.

In order to get good product from an extruder, it is important that the material be completely melted and reasonably uniformly mixed. If there is inadequate shear mixing in the screw, unmelted granules will be present in the extrudate. Proper melting and mixing requires that a suitable screw be used and that adequate back pressure be maintained on the head end of the extruder to get the needed mixing action. This is

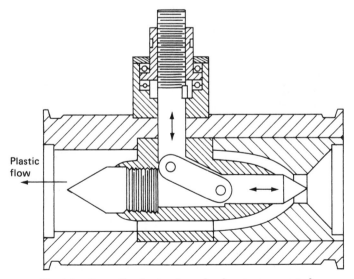

Fig. 2-9a. Streamlined extrusion valve for pressure control.

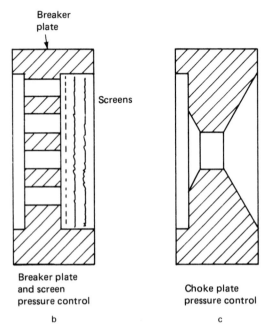

Fig. 2-9b. Breaker plate and screens for pressure control.
Fig. 2-9c. Choke plate for pressure control.

done by the use of screen packs and a breaker plate or by the use of valves or chokes. These three units are illustrated in Fig. 2-9. The pressure drop that the restriction units produce must be added to the pressure drop caused by the die. The screen selection or valve setting is determined by the amount of pressure drop required to produce the desired head pressure in the extruder and the proper inlet pressure to the die for the desired delivery rate.

Another undesirable consequence of incomplete matching of the melting characteristics of the machine to the material is surging. This is a fluctuation in machine output which occurs with a time constant ranging from 1 to 60 sec. Occasionally, longer time constants are possible. The surging results from instability in the melt pool in the screw flights. It usually occurs at output rates which are at a peak for a given machine/ die combination. The pulsations are erratic, and the time constants mentioned are average for a surging condition. In many instances the onset of the condition coincides with a change in the screw speed. The changes in operating conditions in the machine shifts the point of complete melting of the plastic along the screw. If this occurs near or in the metering zone, the result is unstable pumping action with fluctuations in output.

Frequently, it is difficult to stabilize the machine operation once the surging starts. The general tactic is to slow the output and then to build up to the original rates of output. In some instances this will not work, and the machine must be shut down and restarted. The effect of the surging tends to shift the temperatures along the machine barrel, and it is necessary to cool the machine somewhat before restarting. The choice of screw to produce stable output at high rates is one of the most important process decisions in selecting a suitable single-screw extruder. There are computer programs for design of the screws for specific materials and rates, but they are, at best, a general guide to selection. Experience with a specific material and product is essential to finding the optimum screw geometry for a particular product line.

The matching of the extruder output characteristics to the die requires an understanding of the effects of the back pressure on the rate of output, as well as the effect of melt viscosity on pressure drops in the die. Figure 2-10 shows the performance curves for the extruder.

It can be seen that the extruder output at constant screw speed drops from maximum at open discharge to zero at blocked flow. Since the pressure generated at blocked flow can be high enough to damage the machine, this is not a practical condition of operation. The pressure flow-rate curves are different for different screw speeds with increased screw speeds producing higher output. The increases in output are not

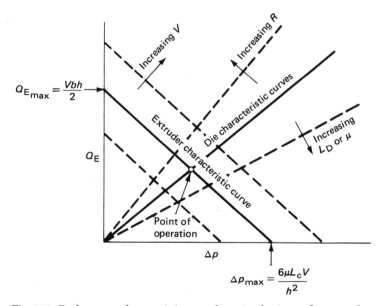

Fig. 2-10. Performance characteristic curve for a simple viscous flow extruder.

proportional to the screw speed. The pumping action changes with the increased shear at higher screw speeds, and the output is less than proportional to screw speed when there is some die restriction. The delivery curves for the dies are shown on the same graph as the delivery curves for the machine in order to find the operating point for the extruder/die combination.

The delivery rate from a die will depend on the pressure drop in the die and the entering pressure from the extruder. The pressure drop in the die will increase with the die opening (diameter, channel depth, channel width, etc.), the channel length, and the apparent viscosity of the material. The viscosity of the resin will depend upon the temperature and the shear rate in the die. The effects of these variables are shown by the die delivery curves. The point of intersection of the die delivery-rate curve with the extruder-operating curve is the operating point for the extruder.

For a given extruder speed and delivery rate the die back pressure may be inadequate to produce properly plasticated melt. This is generally the case. In order to get the proper back pressure in the extruder, it is necessary to increase the pressure drop in the die. This can be done by altering the die geometry—for example, die land-length—or by changing the material viscosity. In practice this would lead to improper shape control of the extrusion, so that it is not a practical approach to proper machine/die balancing. What is generally done, as previously mentioned, is to incorporate a pressure-dropping device such as a screen pack, a choke, or a valve. This will ensure the inlet pressure to the die proper for the desired delivery rate, while the extruder will have a sufficiently high head pressure to properly plasticate the polymer.

For clarity Fig. 2-10 shows straight-line relationships between the pressures and delivery rates. Figure 2-11 shows a more accurate curve which takes into account the fact that the viscosity of the plastic melt is shear sensitive and, consequently, the rates will be pressure and rate sensitive.

This brief exposition on the functioning of a single-screw machine is based on extensive literature in the field. The references cited cover the mechanisms for pumping, melting, and delivery performance in much greater detail. The brief description given is adequate to understand the performance characteristics of single-screw machines in order to design tooling and to operate and troubleshoot an extrusion operation.

The construction of single-screw machines involves the use of components adapted to the stresses imposed and to the control needed for machine operation. The machine size is the main determinant of the drive size, control requirements, and the other elements in the equip-

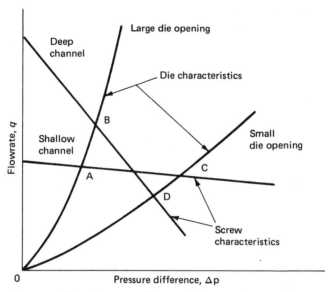

Fig. 2-11. Screw and die performance characteristics curve for a single-screw extruder.

ment. Extruders are nominally sized by the barrel bore or screw diameter. Equipment ranges from a fraction of an inch in diameter to 12 in. (30.48 cm) or larger. The barrel length is the other major size specification, and it is given as the number of diameters long. Typical earlier machines had a length to diameter ratio of 16:1 to 20:1. Current equipment has length to diameter (L/D) ratios from 24:1 to 36:1 and longer. The longer machines have more heat-transfer area and are preferred for more difficult to melt polymers. Table 2-1 shows typical maximum delivery rates for some machines.

The extruder barrels are made from heavy wall alloy steel tubing. They are designed for minimum stretch at their maximum operating pressure [which is 5000 psi (34.45 MPa)] and have a minimum burst pressure of 10,000 psi (68.9 MPa). The end flanges on the barrel can be attached by screws or by shrink and wedge attachments designed to take

Table 2-1. Typical Production Rates for Single-Screw Extruders

1½ in.	(3.81 cm)	24:1 L/D	50–60 lb/hr	(0.0063–0.0076 kg/s)
2 in.	(5.08 cm)	24:1 L/D	90–120 lb/hr	(0.0113–0.0151 kg/s)
2½ in.	(6.35 cm)	24:1 L/D	150–250 lb/hr	(0.0189–0.0315 kg/s)
3½ in.	(8.89 cm)	24:1 L/D	300–400 lb/hr	(0.0378–0.0504 kg/s)
3½ in.	(8.89 cm)	30:1 L/D	350–450 lb/hr	(0.0441–0.0567 kg/s)
4½ in.	(11.43 cm)	24:1 L/D	700–1000 lb/hr	(0.0882–0.126 kg/s)
6 in.	(15.24 cm)	24:1 L/D	1200–1600 lb/hr	(0.1512–0.2016 kg/s)

the burst pressure of the barrels. One end of the barrel is either cut through to form a material feed chute, or it is adapted to be connected to a separate feed chute interposed between the barrel and the gearbox and bearings.

It is now fairly standard to have the barrel lined with a hard cobalt-based alloy, typically X-alloys, to give it abrasion and chemical resistance. There are some barrels still made using alloys which can be hardened by nitriding. The nitrided barrels have shorter lives than the hard-surfaced barrels. Some materials such as vinylidene chloride polymers are very corrosive and require special alloy barrels such as Hastelloy or Z-Nickel. The barrel's precision requirement is that the total out-of-alignment error after all machining operations is less than one-half the screw-barrel clearance. This ranges from 0.002 for 2-in. (5.08-cm) machines to 0.010 for 6-in. (15.24-cm) machines.

In current equipment the barrels are frequently equipped with a vent to devolatilize the polymer. The vent may be plugged and not used when devolatilization is not required. The other details on the barrel itself involve holes suitable for the attachment of thermocouples to sense and control the barrel temperature.

Normal practice in current extruders is to use electrical heat to raise the barrel temperature and either water or air to cool the barrel when required. Various types of heaters such as mica band heaters and rod heaters can be used for heating and different forms of water chests and cooling coils for cooling. One of these is shown in Figs. 2-12. The best current practice makes use of clamp-on aluminum heater–cooler blocks with cast-in rod heaters and cooling lines. The construction of such a unit

Courtesy Waldron Hartig Division, Midland Ross Co.
Fig. 2-12. Extruder barrel showing cast-in resistance heater and housings for air cooling.

4" STARFLEX
HEAT/COOL BAND

Courtesy Glenn Electric Heater
Fig. 2-13. Typical heat/cool temperature control band.

is shown in Fig. 2-13. In the case of the air-cooled machines, the heaters
can be cast aluminum or mica bands, with the cooling supplied by blower
units, and a plenum chamber for each zone of the heater to cool the
barrel by air convection. This heater–cooler arrangement is shown in Fig.
2-14.

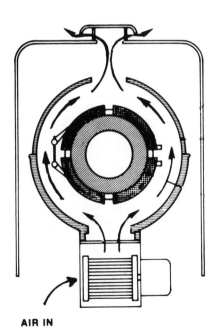

AIR IN

Fig. 2-14. Schematic of cross section of electric heating band and blower cooling
arrangement.

The barrel is divided into a series of zones for heating and cooling, each of which is controlled by a temperature control device. The temperature of the barrel is sensed by the use of thermal sensors such as thermocouples or resistance temperature probes placed in the holes drilled partway into the barrel. Depending on the machine size and the desired melt control, the number of zones will vary from as few as 2 in a sub-inch-size machine, to as many as 12 on a large machine with a high L/D ratio.

The screw is usually made from an alloy steel with high toughness at the melt temperatures for the material. The screw flight tips are usually covered with a hard face alloy, such as Stellite, to improve the wear resistance. The screw configuration is determined by the rate of production required and the polymer to be extruded. Most screws, especially those 2½ in. (6.35 cm) in diameter and over, are cored for a liquid coolant. The shank end of the screw is equipped with a tapped hole that is attached to a rotary union. On most machines the screws have shanks which are the same diameter as the screw. Other machines have the

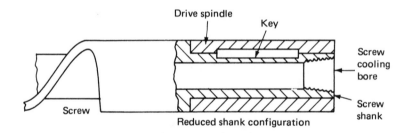

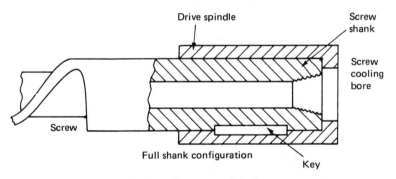

Fig. 2-15. Diagram of shank configurations of single-screw extruder screws.

shank diameter reduced, so that a shoulder is provided against the bearing housing rather than having the rear of the shank rest against a shoulder inside the gearbox spindle. The two configurations are illustrated in Fig. 2-15. The advantage of the full shank diameter screw is that the driving diameter is larger. Driving is done with a key between the screw shank and the spindle. The key is usually fixed, either into the spindle hole, for the full shank diameter screw, or onto the shank, for the reduced shank diameter screw. The key drive is preferred over splines or other more elegant drive connections, for easy maintenance. If the screw is overtorqued, the key should shear off and prevent any further damage. If a spline or other similar drive was used, the spline would twist, making it virtually impossible to extricate the screw for repair without major work on the gearbox.

The main bearing which takes the thrust load from the screw is now frequently incorporated into the gearbox unit. However, it can be in a separate housing attached to the front of the gearbox. The bearings are selected to last 100,000 hr at the extreme thrust load of the screw. This can be a substantial force. Assuming a head pressure of 5000 psi (34.45 MPa) on a 3½-in. (8.89-cm) machine, the direct back force is 50,000 lb (222.5 kN). In addition to this, there is a force exerted on each flight which is transmitted to the screw shank by the drag-flow pressure buildup. Since this total force is difficult to estimate accurately, the rule of thumb is that it is equal to the force generated by the frontal pressure on the screw so that the bearing must withstand 100,000 lb (445 kN) of thrust load for 100,000 hr in the 3½-in. (8.89-cm) screw machine. In well-designed current equipment bearing failures are rare. In some equipment, especially the larger machines, coolant is pumped to the main bearings to keep their temperature down and to increase their life.

For long life, the gearboxes used in single-screw machines are now almost exclusively direct-drive units rather than worm reducers. The gears are either helical or herringbone types for minimum wear at maximum torque delivery. The gearboxes are conservatively rated, since the extruders are generally run for weeks at a time without shutting down. The gearbox ratios are in the range of 6:1 to 12:1 depending on the basic motor speed and the torque requirements of the plastic being run. Typical screw speed ranges are 2–35 rpm, 10–90 rpm, 15–150 rpm, with other ranges available for special cases. In many instances, machines which are designed for custom operations, those that operate on a wide variety of polymers, have change gearboxes, such as the one shown in Fig. 2-16. By switching or changing the gears the gearbox ratio can be changed from a high-speed, moderate-torque range to a slow-speed, high-torque range. It is always advantageous to use the gearbox in the

Fig. 2-16. Change gearbox for single-screw extruder.

range where the drive motor is operating at higher speeds, both for better control and for more efficient motor operation.

The motor drives used on extruders can be variable pitch belt or chain drives, such as the U.S. Varidrive or the PIV drive, or a direct electrical variable drive. The latter can be an eddy current unit, a DC motor driven from a motor generator set or rectifier, an SCR variable speed drive, or a variable frequency AC drive. Most machines are now built with SCR drives. This seems to be the best compromise between cost and ease of control. Some smaller machines are still built with variable-pitch pulley drives because of cost factors. The variable frequency AC drives offer some advantages in stability of operation and speed control, but so far they have made little penetration as a replacement for the SCR drives.

The SCR drives have a control advantage for extruder operation over the variable pulley and straight DC drives because the controls can be remote from the machine. This permits running the entire operation from a central console. The speed can be corrected with a tachometer feedback control and, in addition, special controls can be incorporated to adjust machine speed in response to process requirements. Motor horsepower will be a function of machine size and the material used. For example, the range used for a 2½-in. (6.35-cm) machine can be from 25 to 60 hp (18.65 to 44.76 kW) while for a 3½-in. (8.89-cm) machine, the range is from 40 to 90 hp (29.84 to 67.14 kW).

The controls for both the motor drive and the barrel temperature control are housed in a control cabinet. Sometimes the control cabinet is

Courtesy Welex Co.
Fig. 2-17. Small single-screw extruder with electric heating and blower air cooling. Note blowers beneath machine barrel.

a freestanding unit located near the machine, but the current practice is to mount the control cabinet on the same base as the machine as shown in Fig. 2-17. The temperature controls can be one of a number of different types depending on the requirements of the product and the degree of sophistication of the production system.

Standard practice has been to use a separate pyrometer controller for each zone in the extruder, and separate instruments for controlling die

Bolt flange

Courtesy Egan Co.

Fig. 2-18. Flange bolt die mount for single-screw extruder.

and gate and other temperatures. These instruments use either thermo-couple sensors or resistance temperature sensors. While direct on-off control is occasionally still used, most of the instruments are of the proportioning type and, for the barrel temperature control, are for both heat and cool control. The new equipment may be supplied with a microprocessor-type control which will read each sensor in turn, and feed either heating or cooling to each zone as required. This technology enables closer process control. The subject will be covered in detail in the chapter on control systems.

The head-and-gate section of the extruder is where the extrusion dies are mounted. There are a number of different arrangements possible for the attachment of the die structure to the barrel. Some of these are illustrated in Figs. 2-18, 2-19, and 2-20. The simplest arrangement is to bolt the die directly onto the barrel flange. This requires careful handling of the die unit and some sort of die cart or die-handling unit to move the die away from the machine during the clean-out operations. It is gener-

Split clamp

Split clamp
with
hinged adapter

Courtesy Egan Co.

Fig. 2-19. Split clamp die mount for single-screw extruder.

Hinged gate
with
swing bolts

Courtesy Egan Co.
Fig. 2-20. Swing gate with swing bolt adapter for dies for single-screw extruder.

ally used with large sheet, film. and pipe dies which need the tight attachment to the machine. The hinged gate is a widely used construction with smaller dies. This unit can be secured either with a taper lock clamp, such as the handy clamp, or with swing bolts. Another widely used arrangement is a taper lock assembly using the handy-clamp arrangement, but without the use of a gate hinge. The selection of the head arrangement depends on the size and nature of the die used.

For general purpose use for small shape and pipe dies the gate is bored to take a die-adapter unit. The die adapter is used to attach the dies to the machine, making it convenient to change the dies. As shown in Fig. 2-21, the die adapter is usually drilled to take a melt thermocouple.

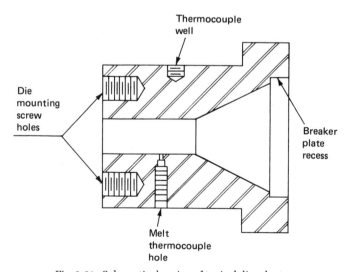

Fig. 2-21. Schematic drawing of typical die adapter.

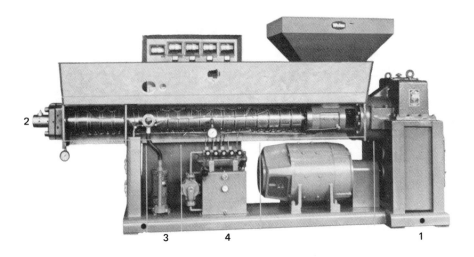

Courtesy Welex Co.
Fig. 2-22. General arrangement of the parts in a single-screw extruder. 1, Drive train; 2, screw and cylinder; 3, venting; 4, heating and cooling.

Figure 2-22 shows the parts of the machine mounted on a base which can be either a casting or a steel-plate fabrication. The barrel is attached to the main bearing which is, in turn, attached to the gearbox. This serves as the back support for the barrel. The front is supported with a pedestal unit that is in light contact with the barrel, both to minimize heat loss to the base and also to permit free expansion of the barrel as it is heated. The motor is also attached to the base with the location depending upon the way it is connected to the gearbox. One arrangement is shown in Figs. 2-23 a and b, with the motor outboard to the extruder, and the connection made through a belt and pulley, usually a timing belt and pulley for positive traction. Another common arrangement is the tuck-under construction shown in Fig. 2-24 where the motor is directly connected to the gearbox and located under the barrel.

The feed hopper is mounted to the feed chute of the machine by screws. The hopper can be a simple unit which will hold about 2 hr worth of material (Fig. 2-25). It is equipped with a gate to stop and start the flow of material, a sight gage to ascertain the material level, and a dump-gate arrangement to empty the hopper when the machine is shut down. Frequently, the hopper has a loader unit attached which moves material from a large container to the hopper, or moves material into the hopper from a central storage silo. Another configuration used for moisture-sensitive materials is a hopper dryer, which will hold enough polymer to ensure that the material entering the feed chute will be

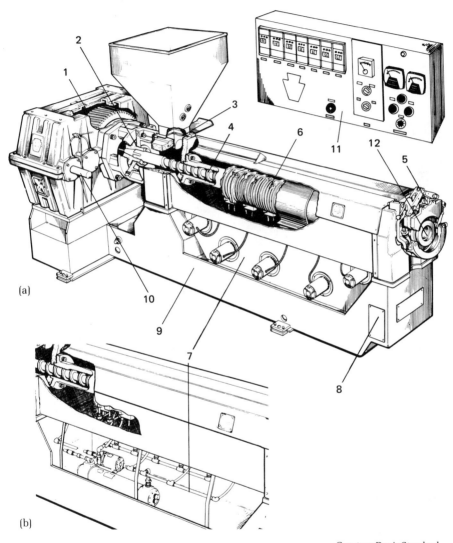

Fig. 2-23. Extruder showing outboard motor arrangement. 1, Gear case; 2, thrust bearing; 3, feed section; 4, barrel and screw; 5, head clamp; 6, barrel heating; 7, cooling systems; 8, electrical wiring; 9, base; 10, drive motors; 11, controls; 12, barrel height.

properly dried. This is as much as 6–8 hr of material for polymers requiring extensive drying.

There are other hopper arrangements used for special cases, such as crammer feeders, metering, and stirred hoppers. These are used in specific products, and they will be discussed further with respect to the processes in the later chapters.

Courtesy Killion Manufacturing Co.
Fig. 2-24. Extruder showing tuck under motor arrangement.

Twin-Screw Extruders

Twin-screw extruders are positive pumping devices which differ significantly in both construction and operation from the single-screw
machines. The theory of operation for the twin-screw machines applies
as well to other multiple-screw machines which have been designed for
use in certain operations. There are a number of different variations
among the twin-screw machines, both as to construction and as to
operational mode.

Twin-screw machines can be differentiated by the screw rotation. The
screws can be corotating or counterrotating and the direction of rotation
can be left hand or right hand relative to the feed-chute location. Each
arrangement has a different effect on the plasticating and conveying
action produced. This discussion will be confined to intermeshing twin-
screw machines which are the type with the positive pressure delivery
characteristics. Figure 2-26 shows the rotational possibilities in the
machine.

Another difference in twin-screw machines is the nature of the
screws. The screws could be constant pitch and constant depth types or

Hopper

Courtesy Egan Co.

Fig. 2-25. Typical hopper unit on a single-screw extruder.

they can be one of the variations shown in Fig. 2-27. These include tapered screws with tapered depth, variable pitch screws of both constant and variable depth, stepped pitch screws with the same or variable depth and constant diameter, and stepped diameter screws with constant and stepped depths as well as different pitches in each screw section. To add to the variety, screws can be equipped with special mixing sections which can mull the material or mix it by controlled reverse flow. In each case the designs represent a particular interpretation of the requirements for processing a specific resin system.

Screw-to-screw clearances are another design variation. In most cases the clearances are not nearly as tight as those used in hydraulic pumps of this general type. The very high viscosity of the polymer melts would cause excessive shear and heat generation and would require very high torque levels. Owing to the high viscosity, the seals that are obtained

Barrel

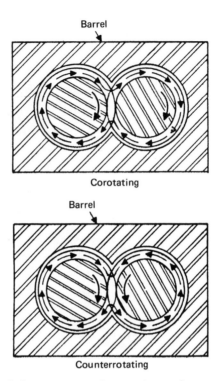

Corotating

Counterrotating

Fig. 2-26. Diagram of the operation of corotating and counterrotating twin-screw extruders.

between the screws with generous clearances (around 1 mm) are good enough to make the machines very positive in their delivery characteristics.

A simplified analysis of the action of the twin-screw extruder is given in Refs. 5 and 6. It indicates that the material movement in any intermeshing twin-screw machine is a result of a chamber or cavity formed between the barrel walls and the intersection of the root of one screw with the land of the other. This cavity progresses down the barrel to the end of the screw where the material is then ejected into the delivery region of the machine. Figure 2-28 illustrates the construction of the screw channel which is analyzed by unrolling in a manner similar to that used for the single-screw machine. Some rather significant differences in operation emerge from the analysis.

One of the results of the positive delivery action of the twin-screw design is that the drag flow and pressure flow which occur do not have a

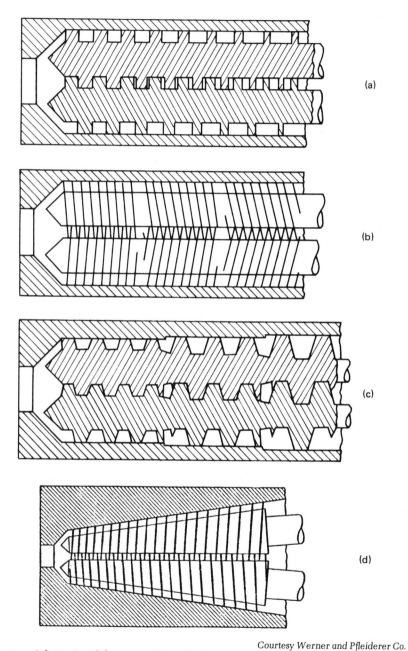

Courtesy Werner and Pfleiderer Co.

Fig. 2-27. Schematics of the construction of twin-screw extruders: (a) Mapré; (b) Pasquetti; (c) Columbo; (d) conical screw with converging axes, Kraus Maffei.

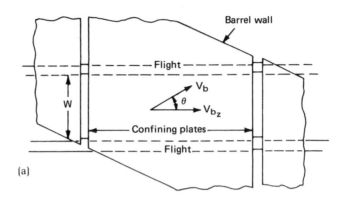

(a)

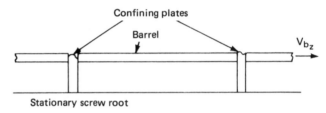

Stationary screw root

(b)   Reprinted from Polymer Engineering & Science 15, 608, August (1975).
Fig. 2-28. Diagram of confining plate model of twin-screw extrusion: (a) overhead view of channel; (b) view parallel to screw flights.

significant effect on the output rate or pressure at the head of the machine. These effects do occur, and it is necessary that they be considered. Figure 2-29 shows the drag flow and pressure flow curves for the twin-screw extruder. In order for the flow to be limited to the progressing cavity delivery, the drag and pressure flows must be additive and equal to the flow of the progressing cavity to produce a dynamic equilibrium condition. The result is that, unlike the single-screw machine, the shear on the material is unaffected by delivery rate and the greatest shear on the material is at the root of the screw instead of at the barrel wall.

The implication drawn from this is that there is relatively little mixing occurring in a twin-screw machine. As shown in Fig. 2-30, this is not the case. Where the land and root meet to make the seal the material is turned over many times. As can be seen from the illustration, the effect is more pronounced in the corotating machines where the effect at the nip is similar to that produced with a two-roll compounding mill. It is,

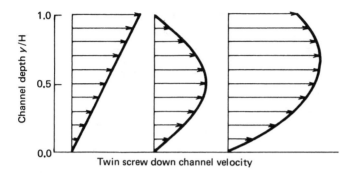

Twin screw down channel velocity

Reprinted from Polymer Engineering & Science **15**, 610, August (1975).
Fig. 2-29. Drag, pressure, and net velocity flow profiles for a twin-screw extruder.

however, much less shear than is produced on the material in a single-screw machine under normal operating conditions.

One of the salient considerations for the use of a twin-screw extruder is that the mixing mechanism ensures that all of the material sees about the same shear history. In addition, the shear produced by the conveying action is much less than that for a single-screw machine and, furthermore, it is controllable. Usually the amount of shear imparted by the pumping screws is less than that desired, so special elements are frequently incorporated (Fig. 2-31). The independent control of the shear and the conveying effects are the major reasons that these machines are

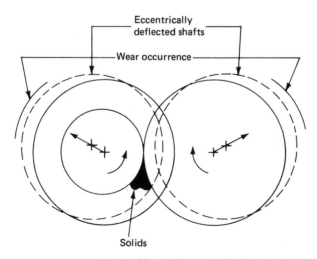

Eccentrically
deflected shafts

Wear occurrence

Solids

Reprinted from Advances in Plastics Technology, **1**(2), Apr. (1981).
Fig. 2-30. Solids grinding in wedge area of counterrotating twin-screw extruder.

Screw engagement		System	Counterrotating		Corotating
Intermeshing	Fully intermeshing	Lengthwise and crosswise closed	1 	Screws 2	Theoretically not possible
		Lengthwise open and crosswise closed	Theoretically not possible 3	4	
		Lengthwise and crosswise open	Theoretically possible but practically not realized 5	Kneading discs 6 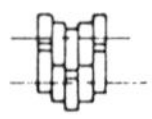	
	Partially intermeshing	Lengthwise open and crosswise closed	7 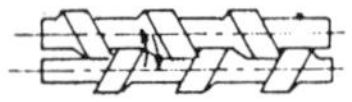	8	Theoretically not possible
		Lengthwise and crosswise open	9A 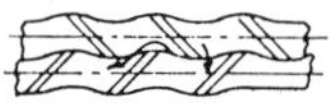	10A 	
			9B 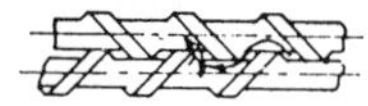	10B 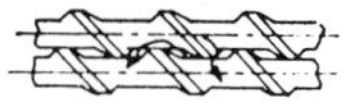	
Not intermeshing	Not intermeshing	Lengthwise and crosswise open	11 	12 	

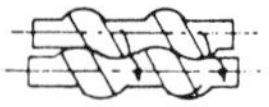

Reprinted from Advances in Plastics Technology, *1(2), Apr. (1981).*
Fig. 2-31. Screw arrangements possible with corotating and counterrotating twin-screw extruders.

used extensively to extrude heat-sensitive and shear-sensitive polymers such as rigid PVC.

The pressure-generating mechanism in the twin-screw machine is the result of the fixed delivery of the softened polymer to the die restriction in the machine. Because of the positive seals in the screws, the pressures generated at the extruder head will be directly related to the pressure drop in the die. If the die is blocked, the pressures can build almost without limit to cause a catastrophic machine failure. The simple interrelation between the pressure and delivery rate does make it much simpler to operate the machine and to design orifices for the dies with the proper pressure drop.

The positive displacement characteristics of the machine changes the manner in which the feed is delivered to the machine. Since the bulk density of the feed is less than that of the melt, it is necessary to compress the material. This can be done by stepwise changes or by continuous changes in the screw flights. The rate of feed of the polymer must be controlled to avoid overfilling the flights. The positive pumping with excessive feed has been known to overpressure the machine and to swell and burst the barrel. The method of avoiding this is to operate the machine with controlled feed matched to the screw rate.

The twin-screw machines are fed with a controlled rate volumetric or weight feeder unit. The feed rate is adjusted to make the extrusion come out at the desired rate. This results in the operation of the machine with incompletely filled channels at the feed end and often at the delivery end of the machine. Unless there is careful adjustment of the feeder rate to the screw speed, the incompletely filled screw-flight condition will exist along the length of the barrel. The amount of material fed must be no more than the amount that it takes to fill the screw flights for the plasticated material. The machine is calibrated for melt delivery, and the feeder is adjusted and checked to deliver just under the maximum amount of plastics needed for full-channel operation.

Unlike the single-screw situation, the heat required to melt the plastic in a twin-screw machine is primarily provided by conduction from the barrel wall and from the screws. This makes the use of heated screws necessary, since the screw surface is a major part of the heat-transfer area available. The screws are cored and heat-transfer oil is pumped through the screws at the necessary temperature for melting the polymer.

The machines used to accomplish the extrusion to make sheet, pipe, and profile, as contrasted with those used for compounding, are usually of the counterrotating screw variety. They cause the minimum amount of shear between the screws, making screw design easier for a specific resin. The design of screws for an application is also made simpler by using experimental screws which are sectional in construction. A particular screw element configuration can be tested before a solid screw is cut. A picture of such a segmented screw is shown in Fig. 2-32. It is used with a corotating screw machine and the elements are keyed onto the drive mandrel which is equipped with a core-cooling arrangement.

The barrels on twin-screw machines have a figure-eight-type bore which is the intersection of two bores matching the screw diameters. These are spaced so that the flight of one screw almost touches the root of the other. Previous practice has been to use nitriding grades of steel to make these barrels and subsequently to harden the surface layer of the steel. Owing to the high levels of wear encountered, the practice is shifting to the use of hard alloys such as X-Alloy materials for liners. The barrels are equipped with heater/cooler units to control the barrel temperature similar to those used with the single-screw machines. Since the machines tend to operate with much lower shear and, consequently,

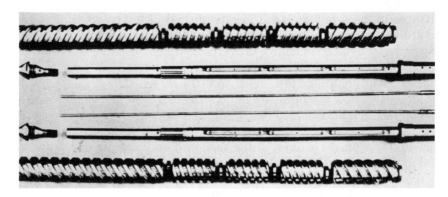

Fig. 2-32. Segmented screws for corotating twin-screw extruder.

generate less shear heating, the preference runs to air-cooled units. Again, the barrels are drilled for sensor units to operate the control instruments.

The barrels are designed to higher barrel operating pressures because of the positive pumping action. The basic barrel structure is less resistant to pressure deformation than the barrel of single-screw machines. Thus the overall thickness of the barrels is greater for twin-screw than for single-screw machines. Unlike the barrels for the single-screw extruders, they are not secured to the gearbox. Most twin-screw machines are designed so that the barrel slides forward away from the gearbox in order to get access to the screws. The machines are equipped with heavy support slide beds to hold the weight of the barrel when it is moved.

The barrel is equipped with a feed chute section which is usually an integral part of the barrel. In some machines, usually those designed for experimental work or for handling a wide variety of materials, the barrel is built in sections. This permits the introduction of a variety of different venting port locations along the barrel, as well as locations where secondary material streams can be introduced. Because of the operating mode of the twin-screw extruders, the location of the vents is very important. A segmented barrel machine can be used to test for the optimum location for the vent. After the appropriate barrel configuration is determined, a unitary barrel can be made. This is desirable from the standpoint of minimizing alignment problems that can cause wear in the bearings and gearbox.

The intermeshing screw geometry places some difficult constraints on the design of thrust bearings and gearboxes for these machines. The centerline distance is fixed by the screw size, and the bearing assembly

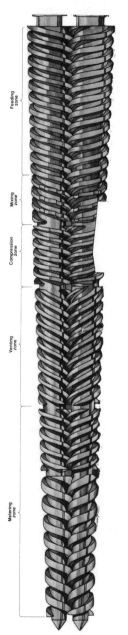

Courtesy Cincinnati Milacron

Fig. 2-33. Screw for conical inclined axis twin-screw extruder.

must be limited in diameter to the centerline distance. In addition, the centerline distance limits the maximum gear pitch size to the diameter of the screws. These problems have been solved in several ways to permit the working life of the extruders to approach those of the single-screw machines.

One solution to increasing the diameter available for both the thrust bearings and the drive gears is to use tapered screws such as those used in the Kraus Maffei machines. Figure 2-33 shows the axis of the screws angled out in order to permit the meshing of the tapered screws. By placing the intermeshing gears some distance behind the barrel, they can be made substantially larger in diameter than the maximum screw size. Placement of the thrust bearings in the same general section of the machine permits the use of larger diameter thrust bearings which can take higher thrust loads. The taper also makes the screws a continuous compression unit that helps the action of the feed.

The gearboxes for most twin-screw machines use extended length gears to permit generation of the necessary torque levels. These extended gears are the ones that drive the screws. The rest of the gear train used is of a more conventional construction. Helical- and spiral-gear configurations can add some to the torque-handling capabilities to the gear drive, but only at the expense of adding to the thrust loads. In the in-line machine configurations, the best way to handle the thrust loads has been by the use of tandem sets of bearings before and after the gears

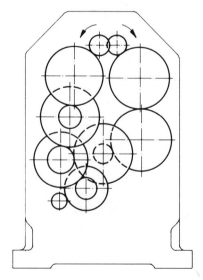

Courtesy American Leistritz
Fig. 2-34. Gearbox configuration for counterrotating twin-screw extruder.

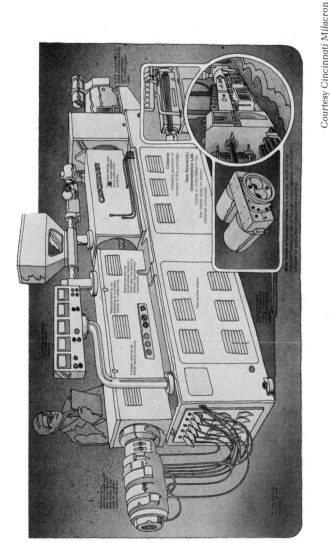

Fig. 2-35. Illustration of twin-screw extruder showing various parts.

Courtesy Cincinnati Milacron

The Milacron Five-Zone Process

Cincinnati Milacron's twin screw design processes material through five zones.

1. The feeding zone provides a large area for rapid heating of the material. The intermeshing screws work like a positive displacement pump to advance the material at a uniform rate regardless of the amount of material in the screw flights.

2. In the mixing zone our design provides a positive mixing action that does not rely on excessive shear between the screw and barrel as in single screw designs.

3. Next the material passes into the compression zone where an effective seal is achieved between the mixing zone and venting zone.

4. In the venting zone an external vacuum is applied to remove any volatile gasses.

5. The metering zone has a special screw flight configuration which advances the melt and compresses it into a steady stream at the output end of the extruder.

Multi-screw extruders offer three kinds of advantages:

A. Direct day-to-day plant operating advantages.

B. Processing advantages that reduce material costs.

C. Machinery advantages that provide better performance and reliability at lower operating costs.

Direct operating advantages

1. **Capability to run lower cost compounds:** This benefit is the result of precise heat control throughout the extruder which allows lower stock temperatures and thus lower cost compounds.

2. **Minimum overweight:** The positive displacement of twin intermeshing screws provides a steady discharge of material. While single screw extruders produce pulsations that can result in 5 to 10 percent overweight, twin screw extruders can cut that figure by more than half.

3. **Long, trouble-free production runs:** The precise control of heat input allows the processor to avoid overheating and degrading of material from excessive friction. This permits long trouble-free runs.

to give sufficient thrust-load resistance to counter the load from the screws. One helpful feature of the twin-screw machines is that the pressure-generating mechanism used produces much less thrust for the same head pressure and screw size than that for the single-screw machine. Figure 2-34 illustrates the configuration for the drive end of an in-line twin-screw extruder.

The delivery end of the twin-screw extruder is somewhat more complicated than that of the single-screw machine. An adapter section must be provided which will converge the figure-eight-shaped discharge from the machine to a single stream. One such unit is shown in Fig. 2-35. The usual practice is to converge the melt into a round plug and then to a breaker plate or ring choke adapter to the die holder, although rectangular entry sections into the die have been used. In most cases direct bolt-on flanges are used to secure the die holders to the machines. However, all of the gate variations described for the single-screw machines are applicable to the twin-screw units.

The heater controls used are similar to those used in the single-screw machines. The use of discrete controllers for each barrel section is the norm, and, generally, the barrel controls are of the proportioning type with SCR outputs on the heating and cooling modes. Melt thermocouples and melt-pressure sensors are used to measure the melt parameters, and often additional probes are used besides the ones at the adapter. The more complex operation of the twin-screw machines is made easier to control by the additional data that are available from upstream sensors.

Just about all of the twin-screw machines are equipped with barrel vents and pumps for devolatizing the melt. This is far more important for the operation of the twin-screw machines because of the positive pumping action of the machines, which can lead to air entrapment and porous extrudates. Standard liquid ring-seal vacuum pumps are used to pump the volatiles. Many twin-screw machines are equipped with injection ports along the barrel to introduce materials such as colorants, plasticizers, and fillers downstream from the main feed. This flexibility allows the handling of materials, including thermoset materials, which cannot be worked on single-screw machines.

The main feed into the barrel is done through the feed chute which is equipped only with a short hopper. The main machine hopper is attached to the machine and arranged to feed the machine through a metering feeder unit. These are usually volumetric feeders, but weight feeders are used in critical operations. To avoid overpressuring the barrel, the output rate from the machine is controlled by the feeder setting.

The drives used on the machines are variable speed units similar to those used on the single-screw extruders. The horsepower requirements are lower for the same output range. For example, one 90-mm (0.28-in.) extruder with a rated output of 900 lb/hr (0.113 kg/s) uses a 60-hp (44.76 kW) drive. The unit has a barrel L/D of 18:1, which is substantially shorter than those typical for the single-screw machines.

The barrel lengths on the twin-screw extruders range from 11:1 to 20:1, with the majority in the range of 16:1 to 18:1. The low helix angle of the screws in the twin-screw machines, usually 10° to 20°, gives much more action of the material at the shorter barrel lengths. The low-helix angle requires much higher average screw speed for the same throughput, so that the speed-reducer section of the gearbox has ratios from 4:1 to 8:1 instead of the greater reductions in the single-screw machines.

The bases for the twin-screw machines are more frequently castings than weldments, probably because of the greater accuracy required for the barrel support slides. The drive, controls, and gearbox are integrated on the base. In most cases the vacuum pumps for the vented ports are also incorporated in the same unit. Freestanding control units are occasionally used, but are not as popular as they once were. Figure 2-35 illustrates one commercial machine with the parts labeled.

The discussion for the twin-screw machines applies as well to other multiple intermeshing screw extruders. For various design reasons, or application requirements, increased capacity is generated by the use of more than two screws. The positive pumping action is retained and most of the theory of operation on the twin-screw machine is valid for the other multiple-screw machines.

General application of the multiple-screw machines is for large-volume production of shapes, of pipe, and for compounding of shear-sensitive materials. They are increasing in use for such applications.

References

1. E. C. Bernhardt, *Processing of Thermoplastic Materials*, Reinhold, New York, 1959, Chap. 4.
2. Z. Tadmor and I. Klein, "The Effect of Design and Operating Conditions on Melting in Plasticating Extruders," *Polymer Engineering & Science* **9**(1), Jan. (1969).
   3. J. T. Lindt, "A Dynamic Melting Model for a Single Screw Extruder," *Polymer Engineering & Science* **16**(4), April (1976).
4. I. Klein, "Predicting the Effect of Screw Wear on the Performance of Plasticating Extruders," *Polymer Engineering & Science* **15**(6), June (1975).
5. C. E. Wyman, "Theoretical Model for Intermeshing Twin Screw Extruders: Axial Velocity Profile for Shallow Channels," *Polymer Engineering & Science* **15**(8), August (1975).
6. K. Eise, S. Jakopin, H. Herrman, and U. Burkhardt, "An Analysis of Twin Screw Extruder Mechanisms," *Advances in Plastics Technology* **1**(2), April (1981).

Chapter 3
Extrusion Dies for Specific Product Lines

Based on the general principles covered in Chapter 1, extrusion dies can be designed to produce stock or special shapes for different end products. The specific dies that will be covered in this chapter are dies for sheet, for film, for pipe and tube, for wire covering, and for profiles. Specialty dies to make other products will be mentioned briefly, with suggested design approaches for these tools based on the design methods for the standard dies.

Sheet Dies

Plastic sheet of varying thickness is a widely used product. It is used directly as a plastics glazing material, or it is formed and fabricated into a wide variety of products ranging from packaging to chemical tanks. The tooling for the process is illustrated in Fig. 3-1. The extruder at the left feeds a stream of plasticated material through an adapter into a wide slit-type die. As is apparent from the diagram, the die opening is adjustable to control the thickness of the sheet. Behind the main die lips is a restrictor bar arrangement which controls the relative flow in different parts of the die to permit adjustments of the thickness across the sheet width. This will be covered in more detail.

The output of the die is fed into a three-roll stack which consists of three counterrotating rollers that operate in a manner similar to a rolling mill. The spacing of the rolls is set to make the sheet of the appropriate thickness. The large surface area of the cooled rolls permits the sheet to be set so that it maintains its size. The sheet is pulled away from the rolls by a set of rubber-covered stripping rolls. Further cooling and other operations are done downstream from the stripping rolls.

Plastics sheets are made in a broad range of thicknesses. The thinnest materials may range down to 0.010 in. (0.25 mm) or less while the thicker material can be up to 1 in. (2.54 cm) in thickness. It is obvious that the die designs will be different for different thickness ranges. Figures 3-2a, b, and c illustrate the internal differences in die cross-section. Figure 3-2a is the cross-section for a thin-sheet die. The shaded region shows the polymer flow volume. The die has a relatively small manifold section and long die lips. There is a secondary balancing manifold located near the exit section of the lips to permit cross-flows to minimize the thickness variations.

Figure 3-2b is the cross-section for a medium-thickness sheet die in the range of 0.100–0.300 in. (0.254–0.762 cm) in thickness. This die is

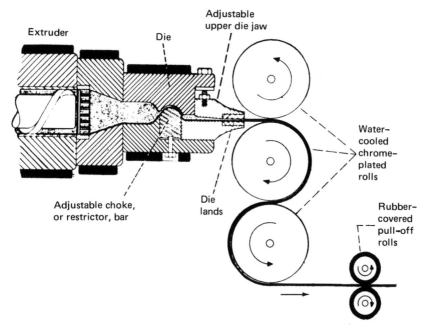

Fig. 3-1. Cross section of sheeting die and take-off unit for sheet.

equipped with a restrictor bar assembly which will control the relative flow to different portions of the die lips by adjustment of the restrictor-bar elements. The secondary manifold used to permit redistribution pressure adjustment is substantially larger than that in the thin sheet die. The L/D of the die lips is smaller since it is less relied on for control than in the thin [0.005–0.1000 in. (0.127–2.54 mm)] sheet die.

Figure 3-2c is the cross-section of the thick [0.250–1.000 in. (0.635–2.54

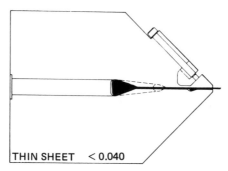

Fig. 3-2a. Cross-section schematic of thin sheet die.

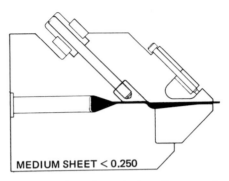

Fig. 3-2b. Cross-section schematic of medium gage sheet die.

cm)] sheet die. In this die the restrictor-bar assembly is much more of a valving unit than in the medium-sheet die, since it is the major controlling effect in obtaining uniform thickness of the sheet across its width. The die lips, while having as large an L/D as possible, are still not an effective enough constriction to act as a flow control for uniformity. The flow control and, consequently, the thickness control are upstream of the die lips. The die lips only act as a final shaping element in the process.

There are two main types of internal die manifold configuration, and these are illustrated in Fig. 3-3. The upper portion of the figure shows a T-type manifold with a constant preland section from the T channel to the die lips section. The lower part of the illustration shows the "coat-hanger" manifold configuration. The main channel of the manifold angles forward toward the die lips to make a triangular preland section. The land length between the main manifold channel and the die lips decreases with distances from the feed point. This results in larger

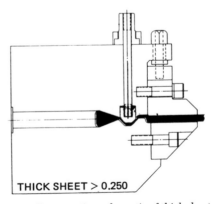

Fig. 3-2c. Cross-section schematic of thick sheet die.

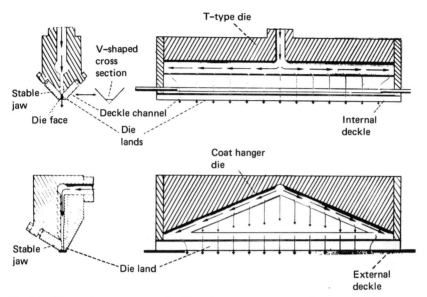

Fig. 3-3. Schematic cross section of a T- and a coat-hanger-type sheet extrusion die. Note internal deckles.

pressure drops on the preland near the feed point and less near the ends of the die. This compensates for the pressure drops in the main manifold feed channel which increases with distance from the feed point. By proper selection of the angle it is possible to obtain a fairly well-balanced flow approaching the die lips and to minimize the adjustment required to obtain uniform sheet thickness.

Figures 3-4a and b show some of the internal construction details of the coat-hanger and T-type dies. It is readily apparent that the dies must be made split along a surface parallel to the plane of the sheet in order to make the required internal detail in the die manifolds. Figure 3-5 shows the construction details in another die and shows clearly the die assembly bolts which hold the die together. It should be pointed out that with the high operating pressures the number and size of the bolts holding the sheet die together are important, and in most cases represent all of the holding power that can be squeezed into the die-manifold area.

Figure 3-5 also shows the die end section which is used to seal the ends of the die lips. Other details worth noting on this illustration are the means by which the heaters are inserted into the die and the method of die lip adjustment, which will be discussed further. This is the Flex-Lip die where the die lips are moved by flexing a thin section of the lips by means of the adjustment bolts. Since this method avoids a joint in the metal in the die lips, there is much less chance for collection of degraded material.

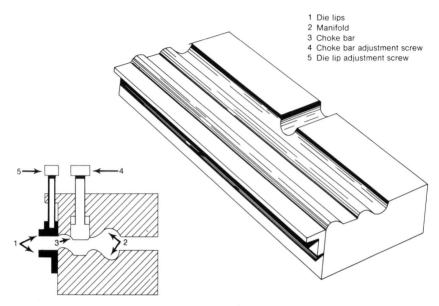

1 Die lips
2 Manifold
3 Choke bar
4 Choke bar adjustment screw
5 Die lip adjustment screw

Fig. 3-4a. Straight manifold sheet die. Lower half and cross section.

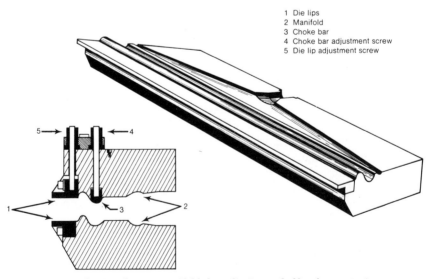

1 Die lips
2 Manifold
3 Choke bar
4 Choke bar adjustment screw
5 Die lip adjustment screw

Fig. 3-4b. Coat-hanger manifold sheet die. Lower half and cross section.

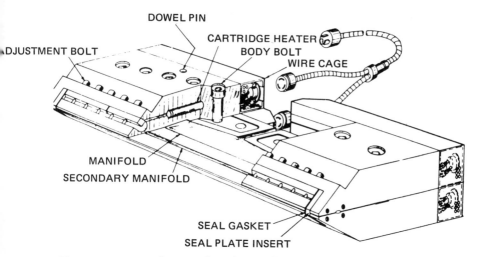

Fig. 3-5. Perspective drawing of coat-hanger sheet die showing internal details.

Figure 3-6 illustrates the construction of a thick sheet die. In addition to making clearer some of the details on the end plate attachment, it shows the manner in which the restrictor-bar adjustment screws are operated. Note that the screws are captivated so that the segments of the restrictor bar can be pulled as well as pushed to make adjustments. The

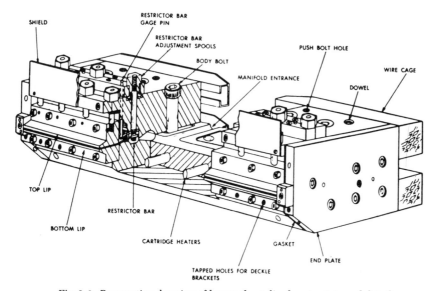

Fig. 3-6. Perspective drawing of heavy sheet die showing internal detail.

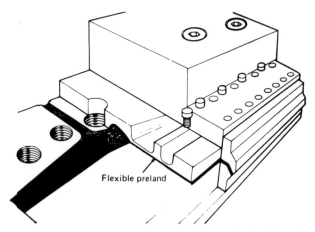

Fig. 3-7. Internal construction of fishtail sheet die with flexible preland.

same adjustments are used on the die lips. This arrangement is necessary because the die pressures on the die sections can change any adjustment unit which is not securely fastened.

One important design development especially important with materials that are readily degraded is shown in Fig. 3-7. This is the use of a flexible preland used in place of the restrictor-bar units. While the range of adjustments of such a device is limited, it is sufficient to be useful in sheet thickness control. It eliminates the interface sections in the restrictor-bar arrangement. This type of die design is used to extrude rigid PVC sheet materials.

Figures 3-8a and b show some additional detail on the construction of the Flex-Lip dies. The length and thickness of the lip section must be carefully calculated to permit sufficient range of motion with the screw adjustment pressure without making the section too thin to hold the melt

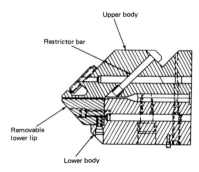

Fig. 3-8a. Flex-strictor die for thick sheet. Removable lower lip accommodates wide range of sheet gages.

Upper body

Lower body

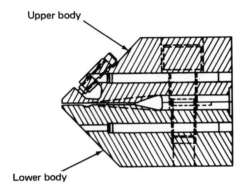

Fig. 3-8b. Flex-Lip die with two-piece body used for thin film or sheet coextrusion.

pressure. A variation of the Flex-Lip construction which adjusts the die opening by altering the bolt length thermally (called the Auto-Flex die) is shown in Fig. 3-9. The use of the die in conjunction with an automatic thickness control system will be discussed in the section on sheet-extrusion systems.

Sheet dies are made in a range of widths suited to the product

LIP ADJUSTING
BOLT

CHOKER BAR
ADJUSTING BOLT

BOLT HEATER BLOCK

LIP ADJUSTING
BOLT HEATER

FLEXIBLE
DIE LIP

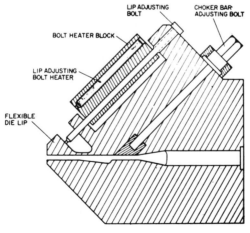

Fig. 3-9. Auto-Flex sheet die. The die opening is controlled electronically by adjusting heat input to die adjusting bolts.

Fig. 3-10. Very wide thin sheet die with Flex-Lip construction. Die is 120 in. (304.8 cm) wide.

requirements. Figure 3-10 pictures a 120-in. (304.8-cm) wide die, which is one of the larger sizes made.

One of the most interesting recent developments in sheet extrusion has been the advent of coextruded sheet. In this material the sheet is made in several layers of polymer of different compositions. The dies used for this type of extrusion are of two general configurations. One configuration is shown in Fig. 3-11 in which there are three independently fed manifold sections with independent flow adjustments in the die. Each of these is fed with a separate extruder and the three melt streams are combined just before exiting from the die.

The other configuration makes use of conventional sheet dies, and the melt streams are brought together in a feed block unit before entering the die. Figure 3-12 is a schematic of the way in which the melt streams are combined in the feed block, and Fig. 3-13 is a diagram of the construction of a feed block unit. The advantages of using standard dies, with their much lower cost and increased flexibility of operation, is obvious. The major problems are with the design of the feed block to give appropriate control of the relative stream flows to give uniform layers of controlled thickness. Recent efforts in this field have resulted in excellent thickness control and uniformity with most sheet-extrudable polymers.

Sheet dies are usually constructed from the 4100 series of semistainless alloys. In most cases the interior of the dies is heavily chrome plated to provide noncontaminating and corrosion-resisting surfaces. Some recent practice has been to use stainless-steel alloys as inserts in the die

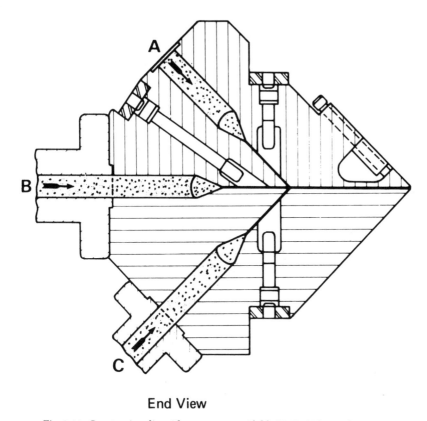

End View

Fig. 3-11. Coextrusion die with separate manifolds (A, B, C) for each resin.

bodies to improve the corrosion resistance, expecially with plastics such as PVC whose degradation products are corrosive. The 17-4 ph alloy which can be hardened to improve its resistance to abrasion is the best choice. Deep nitrided 316 alloy has also been used. The die lips are still generally chrome-plated 4140 steel or a similar semistainless alloy.

Traditionally, conventional machining has been used to make the manifold passages in the sheet dies by contour milling. A recent trend has been to the use of EDM (electric discharge machining) technology where the inner shape is burned into the metal and then later polished. This technique is especially advantageous for making multiple units of the same die since masters can be made of erosion-resistant copper tungsten alloy and used to make several dies before they have to be reworked.

The design of the manifold for a sheet die is a complex problem because of the shear-dependent viscosity of polymers. The degradation

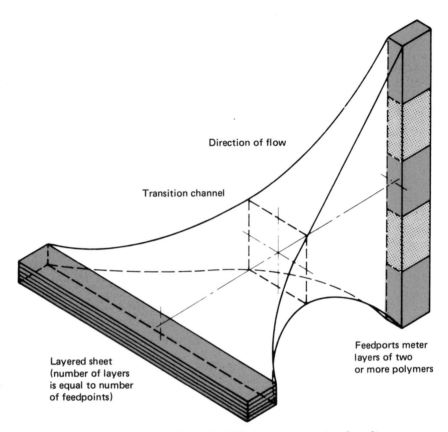

Fig. 3-12. Schematic of flow in feed-block-type coextrusion sheet die.

potential of many materials also makes it important to avoid any dead spots in the flow through the die. There has been a great deal of recent activity in the use of computer-assisted methods to design better die manifolds. One such approach, based on equalizing residence time, is described in Ref. 1. The more general approach is the use of finite-element analysis with simplified Navier–Stokes equations which can predict flow in any type of die. The reference cites some earlier work by McKelvey relating to numerical solutions for sheet die designs.

The designer is required to give reasonably uniform distribution of the material flow in the basic manifold shape. The restrictor bars or preland units enable the flow to be adjusted further to give uniformity in rate across the width of the tool. The final correction is made with the die lip adjustment. It is worth noting that the more effective the previous step,

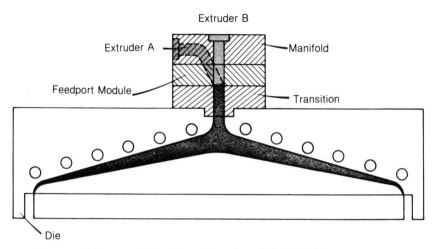

Fig. 3-13. Schematic of fishtail-type sheet die with feed block for coextrusion.

the less adjustment is necessary in the next section of the die. Without effective manifold design it may be impossible to make uniform sheet, no matter what the range of adjustment available in the die lips. The choker bar or restrictor is a correction for the variations in material characteristics and should not be used as a substitute for careful manifold design based on the rheology of the materials to be extruded into sheet.

Film Dies

There are two different production systems employed for the manufacture of plastics film. Film is generally defined as material less than 0.010 in. (0.25 mm) thick. One method, called slot casting, uses dies similar to those used for thin sheet. The second method is the blown-film process, which uses dies to extrude a thin-wall tube of material, and then expands the tube by using air to inflate a captivated bubble of material to make the wall much thinner than that of the original tube.

The slot-casting system is shown schematically in Fig. 3-14. A picture of the end of an actual slot-casting die with the end plate removed is shown in Fig. 3-15. The material exits from the die in a downward direction and is dropped into a chilled, highly polished roll which causes the material to freeze rapidly. The roll is rotating at a fairly high speed relative to the rate of extrusion from the die lips so that the material is drawn down, usually by a ratio of 10:1 or more. In this way, film down to less than 0.0001-in. (0.0025 mm) thickness can be made from reasonable

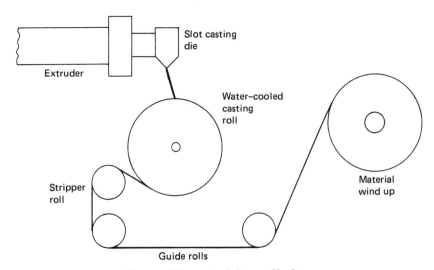

Fig. 3-14. Schematic of slot cast film line.

die slot openings in the range of 0.010 in. (0.25 mm) or greater. One of the materials made by the slot-casting method is polyethylene film because of the rapid quench produces a highly amorphous transparent product. In addition, by the use of suitable additives, the material can be made into a cling wrap for food and similar products.

The bulk of commercial film is made by the blown-film process, shown schematically in Fig. 3-16. The polymer is melted in the extruder and pumped through the film die where it is extruded as a tube in the vertical direction. The tube is captivated by nip rolls, inflated with air from the inside and blown to a bubble about three to six times the original diameter of the extruded tube. In addition, the takeoff equipment is set so that the material is drawn away at three to five times the extrusion speed. Because the resulting material is oriented both in the transverse and machine directions, it gives blown film superior physical strength characteristics.

Several types of die construction are used in blown-film dies. One example is shown in Fig. 3-17, a spiral-flow-type unit. As the plastic flows from the entry point it spirals around the mandrel section of the die. It is clear from the illustration that the land depth between the spiral section and the wall increases as the material progresses through the die. As a result, the distribution around the die periphery is made uniform in order to control the gage of the extruded tube which will, in turn, control the wall uniformity of the blown film.

Courtesy Egan Co.

Fig. 3-15. Picture of slot cast film die with end plate removed.

Another die construction is illustrated in Fig. 3-18. This style is a bottom-entry die with the center mandrel supported by a spider. In order to minimize the possibility that the spider legs, which are barriers in the flow path for the resin, will cause weak spots in the tube (and film), there is a decompression zone after the spider section. In addition, to permit uniformity of melt pressure around the periphery of the die, there are decompression manifolds near the die lips. An additional detail can be seen in this figure with respect to the adjusting screws used to center the die bushing with respect to the spreader or mandrel. This will not

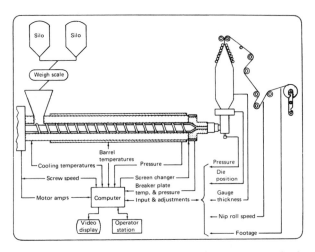

Fig. 3-16. Schematic of a blown film extrusion line with computer monitoring and control.

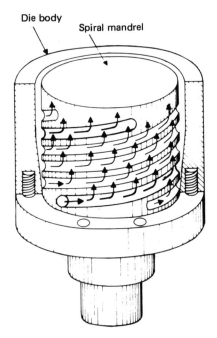

Fig. 3-17. Cutaway schematic of a typical spiral mandrel film extrusion die. Arrows indicate melt flow paths along grooves and across die lands.

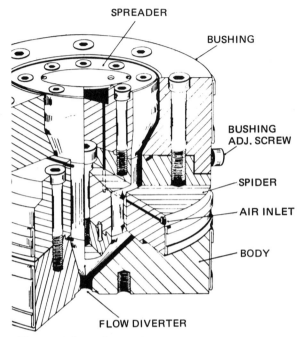

SPREADER

BUSHING

BUSHING
ADJ. SCREW

SPIDER

AIR INLET

BODY

FLOW DIVERTER

Fig. 3-18. Johnson bottom-fed spider-style blown film die.

compensate for out-of-round conditions, but it can be used to correct eccentricities in the film tube.

Illustrated in Fig. 3-19 is a side-fed film die. The mandrel unit is end supported so that there are no spider marks to contend with, but the flow from one side must be balanced to give a uniform rate around the die. This is done partially by the blend path from the entry, and partially by the series of decompression manifolds shown. This type of blown-film die is one of the easiest to construct but it is difficult to maintain uniformity in gage. Careful attention to the blend-section design, based on the melt characteristics of the resin used, will make the flow uniform around the periphery of the die. This will make the requirements on the manifold sections less severe and will permit good gage control. It is apparent that the die is not one that can be run efficiently with a large range of materials and it is also a problem to run at a wide range of rates.

One of the important elements in the blown-film tooling is the cooling of the bubble. In many systems the film is cooled only by the venturi cooling ring which is also used to support and stabilize the bubble. This effect can be seen in Figure 3-20 where both the venturi stabilizing air

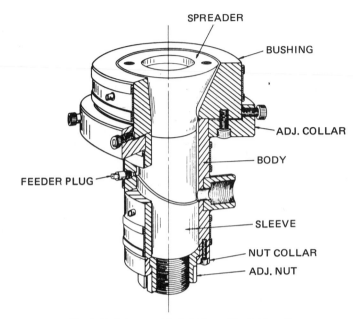

SPREADER

BUSHING

ADJ. COLLAR

BODY

FEEDER PLUG

SLEEVE

NUT COLLAR

ADJ. NUT

Fig. 3-19. Johnson side-fed blown film tubing die.

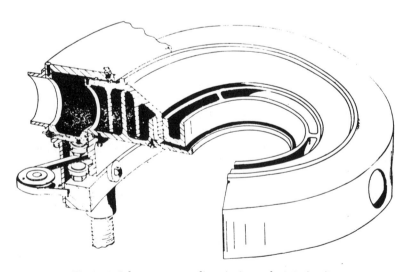

Fig. 3-20. Johnson-type cooling air ring and venturi unit.

flow and the additional external cooling air flow are shown. In Fig. 3-21 the use of an internal air-cooling scheme is shown. Since the air in and out must pass through the die, the die design should be one that has proper access for the incoming and outgoing air. The die is equipped with an extension section that forces the incoming cooler air against the bubble wall to improve the cooling rate. The heated air is then passed out of the bubble through a return air line and vented. Pressure is maintained in the bubble by differential pressure control between the incoming and outgoing air.

Figure 3-22 is a picture of a large blown-film die, the size of which can be judged from the men working on it. A smaller unit is shown in Fig. 3-23. Both figures show the dies mounted on die carts, with the connecting section for attaching the die to the extrusion machine. Each of the die carts is equipped with wheels to enable the units both to be moved and to move back and forth in front of the extruder. Other details that can be seen in the photographs are the heaters which are used to control the die

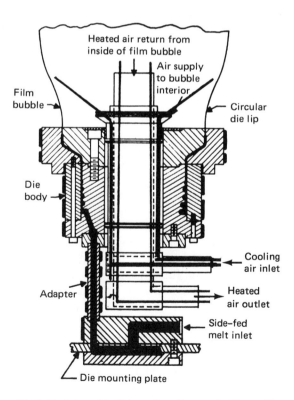

Fig. 3-21. Internal bubble cooling diagram for blown film.

Fig. 3-22. Very large blown film die. Note scale from workmen.

temperatures and the adjusting screws which are used to make the gage uniform.

Figure 3-24 shows a photograph of an internally cooled bubble in operation. The internal cooling unit is seen clearly through the transparent material, with the air deflectors and the return ducts to carry the heated air out of the bubble. The venturi cooling ring is also clearly shown in the picture.

Coextruded film is as important an innovation as the coextruded sheet. By using appropriate combinations of materials, strength, barrier properties, color, and other features of different resins, systems can be combined to make a optimum film. These materials are combined in a coextrusion-film die such as the one shown schematically in Fig. 3-25. Melt streams are fed from the two side ports and are combined at the base of the mandrel section of the die. In the design shown, the combination of the two streams is done well behind the die lips. The die used is a modified end-fed spider-type die. Adjustments of the relative thicknesses of the two coextruded components are done by rate adjustments in the extruders. Another design for coextruded film dies is shown

Courtesy Egan Co.

Fig. 3-23. Small blown film die on cart.

in Fig. 3-26. This type of die is used with materials having widely different melt characteristics. In this design the melt streams are combined just before the die lips. As shown, the design can be modified to make triple coextrusions.

Foamed film is a relatively new form of blown-film product. The films are made from resins with chemical blowing agents incorporated into the material. The die shown is Fig. 3-27 is a spiral-feed unit which has been found to perform well in making foam film. One of the special problems with the foam film is that gage is critical because the expansion of the material as it blows exaggerates the variation in the wall thickness.

Courtesy Egan Co.

Fig. 3-24. Egan's new internal bubble cooling (IBC) system to increase output of low density polyethylene.

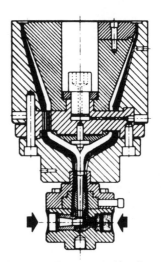

Fig. 3-25. Coextrusion film die.

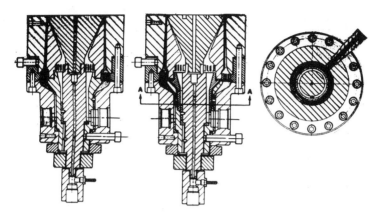

Fig. 3-26. Coextrusion film die for materials with different rheologies.

In the design of film dies, as in the design of sheet dies, one of the important factors is uniform melt distribution and uniform pressure along the die lips. If this condition is met, it is relatively easy to get constant gage in the extruded tube which will be translated into uniform gage in the blown film. The different die bodies described are designed to effect good melt distribution with specific resins. The best all-around design is the bottom-fed spider die since it has the least dependence on the specific material rheology for uniformity in feed. In any event, for maximum production it is essential that the die and manifold feed be adapted to the melt rheology of the resin used.

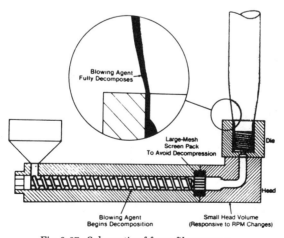

Fig. 3-27. Schematic of foam film process.

Pipe and Tubing Dies

Tubing and pipe are extruded on a line as shown in Fig. 3-28. The die is quite obviously just one element in the production system and it is attached to the end of the extruder. The sizing method used will have an effect on the die construction. One method of sizing is internal mandrel sizing, and a die used in conjunction with this sizing technique is shown in Fig. 3-29. The die holder is a crosshead type and the material flows in from one side and wraps around the pin which forms the inner surface of the tube. It is obvious that this die construction is necessary for the internal mandrel cooling of the sizing unit. It is also much more practical to attach a mandrel to the die if the die holder is a crosshead. In the illustration the method of adjusting the die centering is shown. In this design the die bushing which forms the outside surface of the tube is held to the die holder face with a clamp ring. The bushing itself floats and its position can be set by using the radial adjustment screws. This is done with the ring-clamping screws loosened slightly. After the adjustments are made to make the tube inside diameter and outside diameter concentric, the ring screws are tightened to retain the setting.

The other widely used sizing method for tube and pipe is the vacuum sizer unit. Since this does not place any requirement on the die in terms of sizing equipment, attachment straight-feed dies are used. Figure 3-30 illustrates one construction of a straight-through die. This design is one used to make rigid PVC pipe and tube. The center pin is supported by a spider unit and, as shown, there is no adjustment for centering the pin and bushing. Unlike the crosshead unit, which has its interior open to the

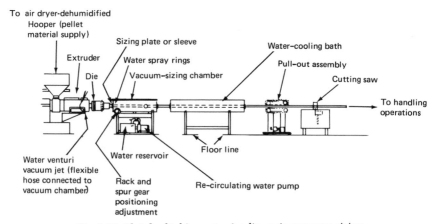

Fig. 3-28. Sketch of tubing extrusion line using vacuum sizing.

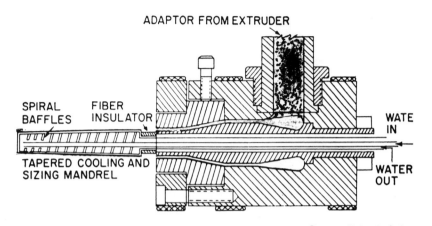

Courtesy Union Carbide Co.
Fig. 3-29. Tubing extrusion die equipped with internal cooling and sizing mandrel.

atmosphere, this die requires an air passage to permit air to enter the interior of the tube as it is extruded. The air inlet runs in through one leg of the spider and then the length of the die pin. When this type of die is used on rigid PVC, it has a choke back-pressure restriction between the die holder and the extruder. The choke is less likely to hang up material and to cause degradation than the usual breaker plate and screen packs.

Figure 3-31 is a photograph of a die similar in construction to that of the die of 3-30. In this unit it can be seen that the die pins are separate units which are screwed into the spider. The pins and bushings are usually made interchangeable so that a range of tubing sizes and wall thicknesses can be made with a limited number of pins and bushings and

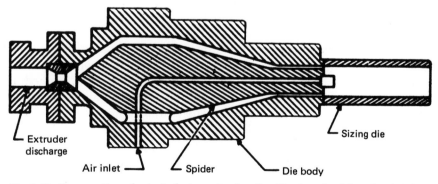

Fig. 3-30. Cross section of a typical pipe extrusion die. The long land length minimizes spider line formation.

Fig. 3-31. Picture of assembly details of a pipe extrusion die.

one die holder and spider unit. Details of the vent hole are readily visible, and the pins used to align the bushing to the spider and die body show one way the dies are constructed.

Figure 3-32 is a drawing of another type of tubing die. In this construction the die holder is a variation of the crosshead die called an offset die. The advantage of this unit is that the product is removed parallel to the machine axis, permitting better plant layout efficiency. This type of unit can be used with either the mandrel-sizing method or the vacuum sizer sleeve. The bushing centering arrangement is similar to that used in the previous illustration. There are two internal details that are interesting. One of these is the flow streamlining ring used to improve the flow of the material around the pin support to make the wall thickness more uniform. This ring can be changed for different materials to compensate for differences in melt rheology. The other internal-flow detail is the flow-restriction section. The restriction produces a suffi-

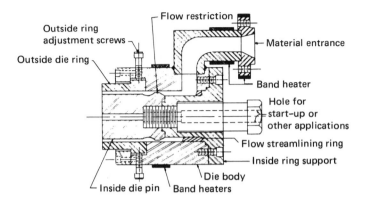

Fig. 3-32. Offset tubing die.

ciently high pressure drop to cause the flow to be made more uniform. Past the restriction there is a manifold ring which permits crossflows to again balance the pressure around the periphery of the tube as it is extruded. These details contribute to wall uniformity and good product dimension control. The die pins and bushings are interchangeable so that the same die holder can be used to make a range of tube diameters and wall thicknesses. This type of die is especially advantageous to use with thin wall tubing, and it is one that has been effective in extruding the cellulosic materials into thin tubing.

Another in-line tubing die is illustrated in Fig. 3-33. The cutaway perspective view shows some detail that was not readily apparent in the previous figures. The centering in this die is done by moving the spider plate with the pin attached, rather than by moving the bushing as in the previous examples. The construction of the spider can be seen clearly in this diagram. Also shown is the air-vent passage entering the central pin via a hole drilled in one leg of the spider. The flow path in this die is a streamline torpedo approach and the flow first diverges and then converges to the exit dimensions. This is done for several reasons. One reason is that this will reduce the effect of the spider leg on the flow. Another reason is that it enables the tool to be used for a wider range of tubing diameters. The largest tube diameter would be approximately that of the spider opening. The manner shown of controlling the die temperature with electric heater bands, operated by a pyrometer sensing the temperature from the thermocouples placed in the die body, is the most widely used.

Figure 3-34 shows another die construction in which there is some internal flow control to minimize the effects of the spider legs on the

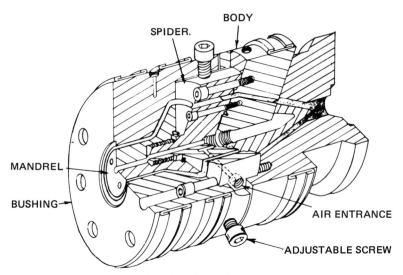

Fig. 3-33. Perspective view of movable pin pipe extrusion die.

flow. In this die construction there is a fairly long straight path past the spider before the material hits the flow restriction. This minimizes the flow disturbance caused by the spider legs. This is further reduced by the pressure drop caused by the flow restriction followed by the open manifold section. This type of die is useful for extruding polyolefin

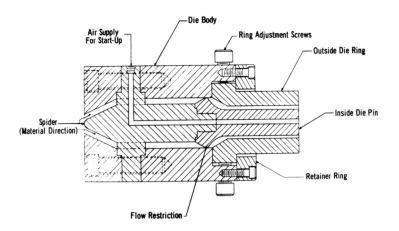

Fig. 3-34. Cross section of straight-through tubing die with restrictor section to minimize spider marks.

materials. The large material inventory in the die makes it less desirable for heat-sensitive materials such as rigid PVC or acetal resins.

Thin-wall tubing made from cellulose acetate, cellulose acetate butyrate, and other cellulosics is frequently difficult to control. The die shown in Fig. 3-35 was designed especially for this application. The main difference in the die is that the pin is undercut for a large portion of its length and that a relatively short land is used to size the tubing. This land is still usually about 10 times the wall thickness, but it represents far less of a pressure drop than would exist without the extended undercut section. The enlarged section will reduce pressure variations around the tube to a minimum so that the wall thickness will remain uniform. It will also damp out any minor short-term variations in the extruder delivery rate and make the wall consistent over the length of the tubing. Its main advantage for the cellulosics is that it inhibits the onset of melt disturbances such as melt fracture and produces very smooth surface tubing with thin walls down to 0.008 in. (0.2 mm) thick.

Smaller dies for extruding tubing are usually mounted directly to the machine either through a die adapter or directly to the gate. For larger size product the dies are too heavy to be supported by the extruder and too heavy for easy handling. The solution is the use of a die cart to support and move the die into position. Such a die cart and die are shown in the photograph in Fig. 3-36. Figure 3-37 shows a close-up view of a tubing die in operation with the material exiting from the die and entering a vacuum sizer. The bushing and bushing centering screws are clearly visible, and the typical neckdown used to make the sizer seal is clearly shown.

Probably the most important consideration in designing tube and pipe

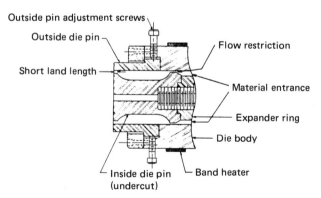

Courtesy Tennessee Eastman Co.
Fig. 3-35. Cross section of die designed for thin wall tubing

Courtesy Gatto Manufacturing Co.
Fig. 3-36. Large pipe die mounted on die cart.

dies is the flow path from the extruder to the die lips. With the internal pin forming the interior wall of the tubing, some means of support is required. This is either end support in the offset and crosshead types or a spider support for the straight-through type. In each case there are flow-path differences that are compensated for by the use of streamline flow deflectors and by the use of restrictions and manifolds. The next important consideration is the adjustment of the die for concentricity. This is usually done by moving either the bushing or the pin by means of centering screws. In some designs, such as those for extruding rigid PVC pipe, the centering is fixed in the die. In most cases, especially with thinner wall product, the centering is essential.

Fig. 3-37. Tubing die in operation showing drawdown into vacuum sizer.

Another factor affecting the decision of which type of die to use is the manner in which the tubing is sized. There are some types of product where the tubing is run directly into a water tank to set the size by controlling puller speed and machine output rate. In most cases the product is sized by the use of a cooling unit for either the inner or outer wall. If an internal cooling mandrel is used, it is necessary to use a crosshead or offset-head die in order to attach the mandrel and to have access to the interior of the die to attach cooling lines to the mandrel. If a vacuum sizer sleeve is used to control the dimensions, this is not necessary and either a straight-through die or one of the offset units can be used. In most cases the straight-through dies are preferable because they have less material inventory in the tool and can be used with thermally sensitive materials with less problems of material degradation. The simpler flow paths also reduce pressure drops and permit more uniform output at higher rates.

While, fundamentally, tubing dies are one of the simpler extrusion dies, they require careful design to obtain optimum performance with different materials. Practical considerations of adapting the units to the machine and to the sizing system as well as to controlling delivery are also necessary and do limit the designer's choice of construction details.

Materials for pipe and tube dies range from carbon steels for polyolefin materials to the 4100 series steels for the general range of materials such as the cellulosics, styrenes, and ABS materials, and stainless-steel or highly chromed dies for the materials with corrosive degradation products such as rigid PVC.

Wire Covering and Other Covering Dies

The extruder is used to cover other materials with an envelope of material. The prime example of this type of product is insulated wire. This product consumes a large amount of raw material to make coated wire which ranges from the very small sizes used in electronic and communication equipment, to the very large sizes used to carry heavy power current. The process is a crosshead extrusion process, and the same general type of tooling is used to crosshead coat other materials, such as wood dowels for towel racks.

There are two basic tool designs for the covering operation, both of which use crosshead or offset head die holders. In one type, illustrated in Fig. 3-38, the material to be covered, wire in this case, passes through the crosshead and a tube is extruded around the wire. The rate of motion of the wire is high enough so that the tube is drawn down until it fits tightly over the wire. The die itself can be recognized as one of the tubing dies described in the previous section.

Figure 3-39 illustrates the other type of covering crosshead. In this unit the plastic material is brought against the wire under pressure inside the die. The mode of operation of this die is quite different from that of the

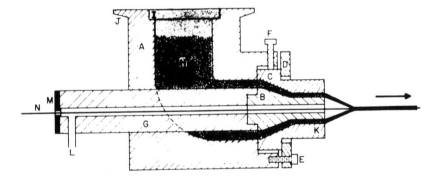

Fig. 3-38. Pressure-type wire covering die. A, Die body, crosshead; B, guider, male die part; C, die, die bushing, female die part; D, die retaining ring; E, die retaining bolt; F, die centering bolt; G, core tubing; H, molten plastic; I, seat for breaker plate; J, ring for attachment to extruder; K, die land; L, vacuum connection; M, wire guide and vacuum seal; N, wire.

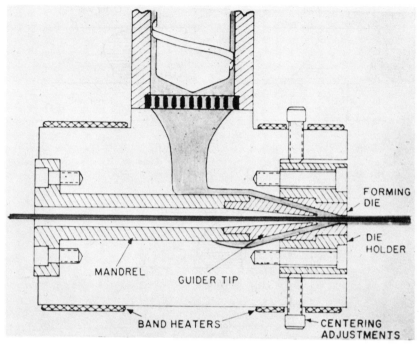

Fig. 3-39. Drawdown-type wire covering die.

tube covering operation of the other crosshead. The plastics material wets the wire, making the traveling wire a factor in pulling the material through the die opening. Extensional flow occurs, and the coating operation is sensitive to the relative rate of motion of the wire and polymer stream. Figure 3-40 from Ref. 2 shows the relative flow caused by drag from the wire and pressure flow from the extrusion pressures. As indicated in the reference, it is possible to have the wire move at such a high speed that the drag or extensional flow dominates and there is actually subatmospheric pressure in the die. When this condition occurs, voids can form between the wire and the insulation leading to a defective product. In some instances it may be possible to correct the condition by applying a vacuum to the entering wire guide but, in general, the requirements are to limit the wire speed or to generate higher die pressures to be certain that there is a positive pressure of plastic melt on the wire.

Crossheads for wire coating are one of the problems that has been approached by the use of finite-element analysis to determine optimum die configuration for high production. There have been a number of papers on the subject (Refs. 3 and 4). In Ref. 4 the design of a die to

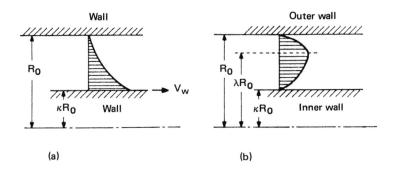

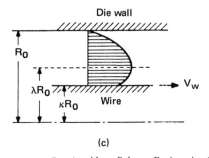

Reprinted from Polymer Engineering & Science *18, 1028, Oct. (1978).*
Fig. 3-40. Schematic of velocity profiles in (a) drag flow; (b) pressure-driven flow; (c) combined drag and pressure-driven flow.

produce pressure flow is shown, and Fig. 3-41 from this article shows both the pressure relationships and streamlines resulting from a finite-element-method design. The superior performance of such dies justifies the analytical design.

Coextrusion coating of wire is done to provide special insulation. An example of a die to make a foam and solid resin wire jacket is shown in Fig. 3-42, with the internal path structure shown in Fig. 3-43. This die is a good illustration of the complexity of paths for the material required in coextrusion dies in general and will be reviewed in conjunction with the section on coextrusion.

Covering operations are not limited to wire. A number of other products such as towel bars and metal tubing are crosshead extruded as shown in Fig. 3-44. Unlike the wire, there is a much more complex problem involved in getting a uniform coating around stiff shapes, especially if they are not round. The melt effects in a wire die tend to make the wire self-centering, especially in the smaller diameters. If the wire moves closer to one wall of the bushing, the pressure generated

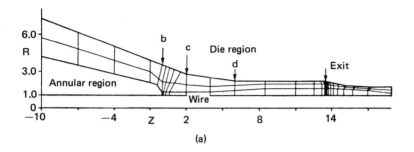

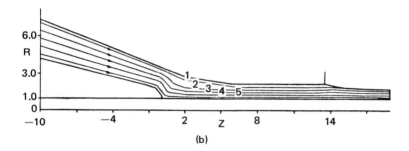

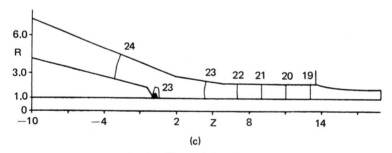

Reprinted from Polymer Engineering & Science 18, 420, April (1978).
Fig. 3-41. (a) Finite-element grid, (b) streamlines, and (c) pressure contours for wire-covering die designed by finite-element analysis.

becomes higher on that side and lower on the opposite side due to the change in L/D of the channels. The unbalanced pressure condition will deflect in the centered position to balance the pressure. If the product coated is stiff, however, it requires much greater forces than can be generated this way to displace an off-center part back to a centered location, so that covering stiff sections, and especially covering nonrounds, complicates the centering problems.

Dies for wire coating are made from essentially the same steels as

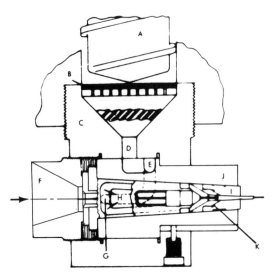

Fig. 3-42. Coextrusion wire-covering die. A, Stock extrusion screw; B, straining screen and backing plate; C, extrusion head; D, feed port; E, channel legs for foamable plastic; F, conductive core; G, bore (one or two) to inner channel legs; H, inner channel legs; I, die; J, auxiliary extruder (not shown); K, annular manifold for outer skin melt.

those used for tubing and sheet dies, except for the portions involved in the wire-guiding operation. The use of the 4100 series semistainless steels is common, and for corrosive resins the use of 300 series stainless steel and the 17-4 ph materials is normal. The guide pins are usually made from high-wear tungsten tool steels because of the high wear rate produced by the wire passing through the die. It is not uncommon to find

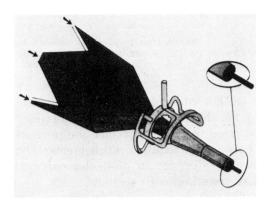

Fig. 3-43. Flow paths for coextrusion wire covering die.

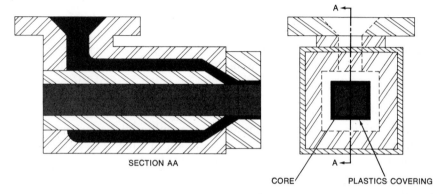

SECTION AA

CORE   A   PLASTICS COVERING

Reprinted from Plastics Machinery & Equipment 7(2), 44, Feb. (1978).©
Fig. 3-44. Crosshead die used to cover towel bar.

the use of ceramic and sapphire bushings in the wire guide sections of wire-covering tools.

Probably the most significant design areas in wire coating and other covering dies is the design of the internal flow paths to make a uniform coating on the product. The crossheads introduce flow-complexity problems which can be solved by design calculations tempered by experience as to what is necessary to ensure uniform pressure around the wire in order to produce a uniform wall thickness of insulation. Another problem that can occur at high line speeds—usually over 1800 ft/min (550 m/min)—is the reduction in internal die pressure caused by drag or extensional flow. This problem can be solved either by changing the die shape or by adjusting the relative rates of wire movement and material flow. The design of good quality high-output wire and other covering dies is fairly straightforward given the state of the art.

Rod and Profile Dies

The rod and profile dies are grouped because they both represent simple shaping operations done in the extrusion process, primarily in the extrusion die. Rods are one of the standard stock shapes produced in plastic along with strip, square bars, thick narrow slab stock, and other shapes generally used in machining and other fabrication operations. The principles used in designing profile dies apply equally to the stock shapes, and the main difference is one of end use rather than of die and process.

There are two general categories of dies used for profiles in terms of general construction. These are plate dies and streamline entry dies. The first is illustrated in Fig. 3-45 and shows that the shaping orifice is in a flat plate attached to a die holder. It is evident that there is a great deal of stagnation in the region behind the die. For some materials, such as the polyolefins, which are thermally stable, this is not a serious problem. On the other hand, thermally sensitive materials, such as rigid PVC, cannot be run for any significant length of time in plate dies. The degraded material which forms in the stagnation regions will start to release stained and decomposed resin and the result will be an unusable product. Figure 3-46 illustrates a streamlined die which has a transition section that matches the flow from the extruder head diameter to the product shape. This type of tool construction is used to prevent stagnation of material in the die.

Round rod can be extruded in either type of die, but the normal practice is to use a die with a conical approach die or a more streamlined unit as shown in the illustrations of Fig. 3-47. The steeper angles are used with materials such as rigid PVC and acrylic resins, and the shallower angles are used for polyethylene. Probably the major consideration in extruding round rod is the die swell which makes the rod diameter larger than the die opening. This is compensated for by adding to the drawdown and the rod can be accurately sized. The round rod is a one-dimensional flow extrusion, which means that the flow is all in the machine direction and is one of the easiest shapes to extrude. Other stock shapes such as squares and strips represent more complicated tooling.

The die to make a square section is a good one to use to illustrate the

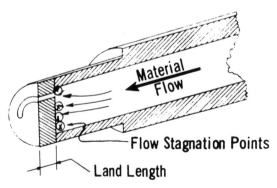

Courtesy Tennessee Eastman Co.
Fig. 3-45. Plate-type profile extrusion die.

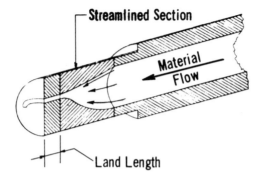

Courtesy Tennessee Eastman Co.
Fig. 3-46. Streamline profile extrusion die.

die construction required to get good product shape. The square section is a two-dimensional flow problem which indicates that in addition to flow in the machine direction there is some flow across the extruded section. In some instances the approach section of the tool is a major influence on the profile, and with a simple conical approach, a die with a square opening can produce an essentially round part. The die must be corrected by adjustment of the die channels to make the product square. Figure 3-48 shows several versions of a die to extrude a square shape.

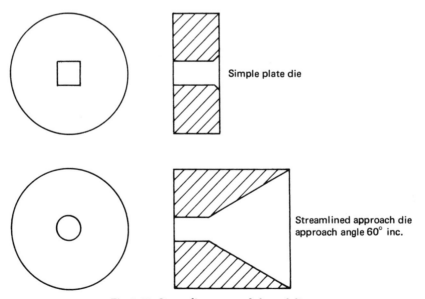

Fig. 3-47. Streamline approach for rod die.

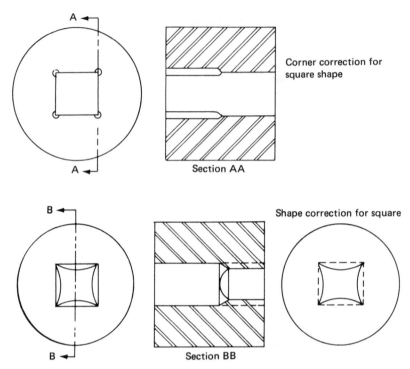

Fig. 3-48. Shape corrections needed in profile die for square shape.

The upper illustration shows the method of producing a square corner in the part by modification of the channel length, and the lower illustration shows a modification of the channel shape. If neither correction is applied, the part will come out of the die with heavily rounded corners as a result of the extra drag in the corners of the die. By reducing the land length in the corners, the drag is reduced and the corners fill out to make the shape square. The alternative method is to exaggerate the shape by deepening the corners so that additional material flows in the corners to make the part square.

The two different approaches can make the part but there are substantial differences in operation and correction requirements. In the case of the reshaping of the die it is a cut-and-try approach which depends to a great deal on the experience of the designer and toolmaker as to exactly what the shape must be. In addition, since the materials are very shear-sensitive as to viscosity, the range of operating rates is closely limited to the one for which the corrections were made. In the case of the land length, correction estimates can be made on the length of land removed in the corner based on pressure-drop calculations so that flow can be increased to the correct amount. In addition, the correction can be changed in a fairly predictable manner since the pressure drop is proportional to the land length. In the case of the shape-change approach, one of the problems is that the pressure is a cubic function of the channel depth making the adjustment very sensitive to the size of the effective channel. In operation the land length corrected die will generally have a significantly wider operating output range.

Another stock profile shape which is widely used is a plastics strip. This shape is useful in demonstrating the problems in defining and producing a profile. Figure 3-49 shows two different methods of correcting a strip die for the corner rounding which occurs as a result of the drag produced by the channel restriction. As in the case of the square shape, the corner can be filled out by reshaping the die orifice by exaggerating the corners, or by reducing the effective land length in the corners. It is also true in this case, as with the square, that the better method is the reduction of the corner land length since it is more predictable and is less sensitive to the amount of adjustment.

There is another effect that must be accounted for in the strip as it is extruded from a streamline die. This is the tendency for the center of the strip to be thicker than the ends. This is caused by the fact that the conical blend makes the material path and the restriction on the flow less in the

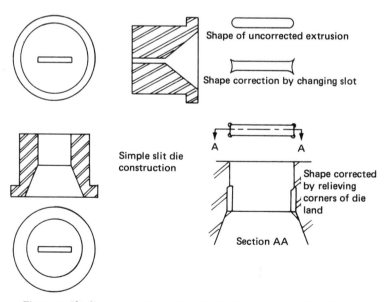

Fig. 3-49. Slit die construction with methods of corner correction of shape.

center of the die than at the edges. This is compensated for by changing the land length as shown in Fig. 3-50 which shows a perspective cutaway view of a strip die. The curved approach section of the land is adjusted to make the pressure drop uniform along the strip width. With these corrections the dimensions of the strip can be controlled except for die swell effects.

The die swell is a characteristic of the material which results in the material expanding from the die dimensions as it exits from the die orifice. The effect is shown in the illustration of Fig. 3-59. It is evident that the land length has an effect on the degree of swell, as has the material formulation. Using long land lengths will minimize the effect, and external lubricants in the resin will minimize the drag of the die and reduce the swell. The amount of swell is a function of the shear rate given a specific formulation and die. As a result, in the case of the strip, the swell percentage will be low for the long dimension of the strip as compared with the swell for the thickness. As a result, a rule of thumb for correcting is that the same absolute value of swell will occur around the section, adding the same amount to both width and thickness. Since this sort of change in relative dimension cannot be removed by the draw-down of the section, it is necessary to compensate for the die swell by changing the length-to-width ratio if the proportions are important to the use of the strip.

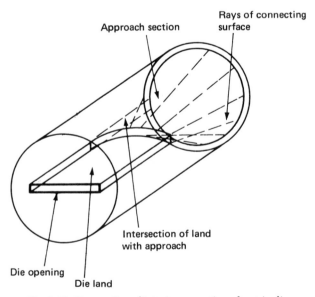

Fig. 3-50. Perspective of interior correction of a strip die.

The example of stock shapes brought out the basics in shape control
for profiles. Strips with low thickness compared to width are one-
dimensional flow dies, while a square is a two-dimensional flow die. The
difference is related to whether there is crossflow in the die shape
perpendicular to the machine direction. Strip die performance can be
predicted on a theoretical basis,[5] and most simple profiles are made up of
collections of strips attached to each other. If the strip sections are all of
the same thickness, then the only complicating factors are the interactions of the strips. The manner of dealing with these is covered in Ref. 6,
which covers in detail the profile extrusion techniques.

There is one group of two-dimensional flow dies that has predictable
enough flow so that dies can be designed with reasonably forecastable
performance. These are dies made up of thick and thin strips. Such a die
for an unequal thickness leg angle is shown in Fig. 3-51. The die land
length for each leg can be calculated. In the case of the thicker leg the
controlling land is a restriction placed in back of the die lips. In the case
of the thinner leg the controlling restriction is the shortened die land
produced by opening the slot back of the exit section. The result is that
the flows from each section of the die will be made in proportion to the
required thickness. At the intersection between the two legs of the angle
a splitter is used to isolate the melt streams for each leg from each other
until a few die thicknesses from the exit point to permit the sections to be

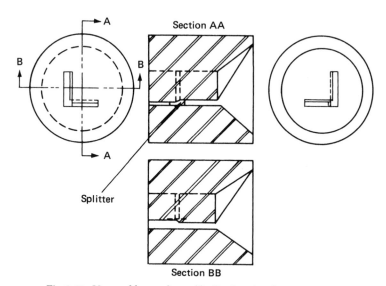

Fig. 3-51. Unequal leg angle profile die showing shape corrections.

joined. The result is good predictability of the shape and reasonable corrections to make the part to dimension.

It is beyond the scope of this presentation to go into more detail on the profile die design. Each of the cases of shape type requires a different manner of determining the relative flow in each part of the profile. In addition, each material used has a melt rheology that must be considered in the die design. Some shapes are essentially impossible to extrude with some materials. Polyethylene and the other polyolefins are among the most difficult to shape, and rigid and flexible PVC compounds are among the easiest to shape. This is shown in the comparative chart Fig. 3-52. Use of this chart is a useful guide to the profile extrusion as explained in Refs. 7 and 8.

Before completing the discussion on profile dies, it is desirable to discuss further the question of approach sections or streamlining, since this can have a substantial effect on the shape. One aspect of this is the transition from the round shape of the material as it leaves the machine to the part shape. This can be done by using transition-type geometry to get a uniform change in section. The other aspect is the actual flow pattern that polymers make entering dies from a die adapter. Figure 3-53 from Ref. 9 shows that the convergence is highly dependent on the polymer characteristics. This makes true streamlining design a rather complex job. One of the pragmatic approaches to this is illustrated in Fig. 3-54. A plate die, with the proper corrected orifice, is attached to a die holder, and the plastics material is run through the die for a period of

Fig. 3-52. Table of Extrudeability Characteristics for Profile Extrusion[a]

Material	Thixotropy	Melt Viscosity	Temperature Setting Range	Frictional Heat	Melt Elasticity
Rigid PVC	1	1	3	4	4
Flexible PVC	2	2	3	2	3
Acrylic	3	2	3	3	6
Modified PPO	2	3	2	2	5
ABS	3	3	3	3	4
Impact styrene	4	4	3	2	4
Polycarbonate	4	3	4	4	4
Thermoplastic polyester	4	4	2	3	4
Polypropylene	6	4	3	2	4
High-density polypropylene	6	5	5	2	2
Polyethylene	7	5	6	2	4
Polyamides	8-9	7	3	2	3

[a]Numbers represent the relative difficulty involved in extruding profiles of the materials. A low number, on a scale of 1 to 10, indicates that particular characteristic of the material makes it relatively easy to form.

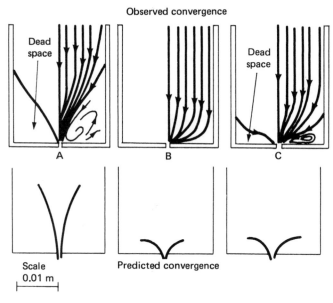

Observed convergence

A   B   C

Scale 0.01 m   Predicted convergence

A   Low density polythene (MFI 2–0) at 190°C (374°F)

B   Extrusion grade polypropylene at 230°C (446°F)

C   Moulding grade acrylic at 230°C (446°F)

Reprinted from Polymer Engineering & Science **12**, 69, Jan. (1972).

Fig. 3-53. Natural die entry flow profile for several materials at a shear rate of 300 sec^{-1}.

time. In the case of a thermally sensitive material such as rigid PVC, the stagnant portions behind the plate die will discolor due to extended heat exposure. In the case of more stable material, a color-changing dye or pigment which is heat sensitive can be added to the melt. In either case, the cull in the die will clearly show the streamline approach to the orifice. By sectioning the cull to get the shape, and reproducing it in steel or other die material, the proper streamline for the approach section can be obtained by experiment.

Dies for profile extrusion are made using conventional as well as EDM methods. The steels range from carbon steels for low-cost short-run plate dies, to the use of 4100 series semistainless steels for most materials with both plate and streamline die types; 300 series stainless steels and 17-4 ph steels are also used frequently for long runs with corrosive polymers. In almost all cases when the carbon steels and the 4100 steels are used for long runs, the dies are plated with either nickel or chrome to maintain the surface quality of the steel. Increased roughness due to thermal corrosion can alter the die performance and lead to changes in the profile shape.

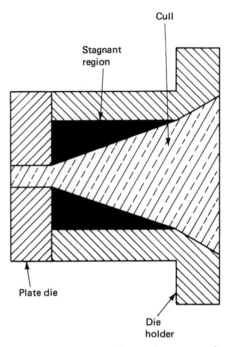

Fig. 3-54. Die cull formation that produces streamline flow.

The subject of profile dies is one of the more complex in extrusion. In addition to the use of careful consideration of flow in the dies on an analytical basis, there is a substantial need for experience in the prediction of die performance, especially for two-dimensional flow dies. The usual design process includes the best shape for the flow based on the material rheology and experience, with the anticipation that the die will have to be corrected after testing to make the shape conform to the drawings. Materials are critical since the rheology of many of the common materials is such that good profile extrusion is virtually impossible with complex shapes and intractable resins.

Figures 3-60, 3-55, and 3-56 show several commercial dies to indicate the construction. The channel die in 3-60 is used for rigid PVC and the part size is about ½ in. × ½ in. (1.27 cm × 1.27 cm). The corrections necessary to make the part with a uniform wall are clearly visible. Figure 3-55 shows the much larger die for a fluorescent fixture cover with heaters in place on the extruder. The manifold to spread the material is visible behind the die plate, and the machine mounting is done using bolts into an adapter ring. The complex die with a ribbed section that makes the part shown in Fig. 3-57 is shown in Fig. 3-56. The plate for

Fig. 3-55. Extrusion die for lighting fixture wrap extrusion.

Fig. 3-56. Complex extrusion die open to show internal details.

Fig. 3-57. Vacuum sizing fixture for and part made by the die in Fig. 3-56.

supporting the cores is separate and stacked with the adapter plate which shapes the incoming melt flow, and is fronted with the main die plate. From these illustrations it is clear that the dies can range from fairly simple constructions, such as the channel die, to complex machining jobs, such as the cover shown in Fig. 3-57.

Specialty Dies

There are many dies which cannot be classified in any of the categories that have been discussed so far. They make unique products and have construction features which differ in many ways from conventional dies, even where the design principles are similar. Examples of this are dies used to make plastics netting. These tools are fundamentally strand or rod dies with multiple orifices. Making the orifices move relative to each other results in an interlocked set of strands to make the netting. Another example is the manufacture of a composite plastics–metal profile, such as the one shown in Fig. 3-58. Here a plastics shape envelopes a metal profile. The part is for a corona discharge static eliminator. It is made by a crosshead extrusion die, but the die is one which has to be made with internal flow control to avoid deflecting the part, and with supports in the crosshead to support the metal section without interference with the material flow.

Fig. 3-58. Extrusion-covered aluminum channel.

Other examples, such as coextrusion dies, can be added to the list of specialty dies. Some of these will be discussed under the systems descriptions since the product can only be made using cooperating equipment. The general flow-path design is similar to that of the other classes of dies which have been covered. These products and tools represent the most advanced concepts in extrusion, which make the most interesting products.

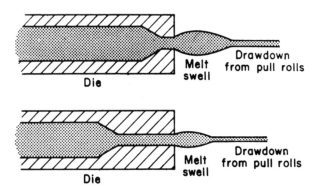

Fig. 3-59. Diagram showing effect of land length on die swell.

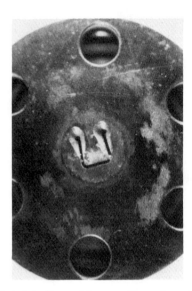

Fig. 3-60. Corrected plate die used for extruding a channel from rigid PVC.

Summary

This section has been concerned with the design of the dies used with some of the major types of extrusion systems built. Each of the product types has its requirements on the flow passages in the dies, material characteristics matching, and construction, which are tied to the shape made and the material processed. One of the main factors considered in die design is the rheology or melt characteristics of the resins. In some instances, such as the production of complex profiles, selection of the resin may preclude a successful die. The designs must reflect the handling characteristics of the materials.

Die construction is usually done by both conventional machining and by the use of EDM techniques, which have become widely used recently because of the ease of making the complex shapes required for the die flow paths. The materials used for the die construction range from carbon tool steels, through the chrome moly and chrome vanadium semistainless steels, to the high nickel and chrome stainless steels. The material choices are usually based on the corrosivity of the degradation products of the polymers. A secondary consideration is the thermal conductivity of the metals which can affect the die performance.

The general principles of die design and construction discussed should be a basis for an intelligent approach to die design and selection.

References

1. Y. Matsubara, "Design of Coat Hanger Sheeting Dies Based on Ratio of Residence Times in Manifold and Slot," *Polymer Engineering & Science* **20**(4), July (1980).
2. C. D. Han and D. Rao, "Studies on Wire Coating Extrusion," *Polymer Engineering & Science* **18**(13), Oct. (1978).
3. C. Gutfinger, E. Boyer, and Z. Tadmore, "An Analysis of a Crosshead Die with the Flow Analysis Network (FAN) Method," *Polymer Engineering & Science* **15**(5), May (1975).
4. B. Caswell and R. I. Tanner, "Wirecoating Die Design Using Finite Element Analysis," *Polymer Engineering & Science* **18**(5), April (1978).
5. E. C. Bernhardt, *Processing of Thermoplastic Materials*, Reinhold, New York, 1959, p. 248 et seq.
6. *Report on Plastics Profile Tooling and Technology*, S. Levy P.E. & Associates, P.O. Box 9502, Fresno, California 93792, May, 1979.
7. S. Levy, "Maximizing Flow through Profile Dies?," *Plastics Machinery & Equipment* **7**(4), April (1978).
8. S. Levy, "Melt Rheology: Its Effect on Profile Extrusion Dies," *Plastics Machinery & Equipment* **7**(9), Sept. (1978).
9. F. N. Cogswell, "Converging Flow of Polymer Melts in Extrusion Dies," *Polymer Engineering & Science* **12**(1), Jan. (1972).

Chapter 4
Controlling the Extrusion Process

Effective operation of the extrusion process requires that the machine and tool operating conditions be maintained in the range determined by the material characteristics and the die performance characteristics. This requires control over the process parameters such as temperature, pressure, machine speeds, and the relative speeds of the auxiliary downstream equipment. The effectiveness of this control will be the major determining factor in the control of product dimensions and uniformity. It will also strongly influence the rates at which acceptable product can be manufactured.

The temperature of the plastics melt is one of the most important parameters in the extrusion process. It is determined by the amount of heat transferred to and from the melt through the extruder barrel wall and by the amount of heat generated by the shear of the material in the machine which generates frictional heat. The melt temperature is generally measured with a melt thermocouple inserted into the die holder. The probe end is extended into the melt stream (Fig. 4-1). The temperature which is measured is used as an overall process-determining datum since it cannot be directly controlled. What is controlled is the temperature of the extruder barrel and the die and adapter zones. These temperatures determine the heat-transfer conditions in the machine, and are one of the determining factors in the melt temperature.

The barrel and die temperatures are controlled by means of a pyrometer controller which senses the temperature of the machine section. It is connected to, and operates a heat input or heat extraction unit to deliver heat to, or remove heat from, the machine section. One of the most common types of controllers used is a thermocouple sensor attached either to a sensitive galvanometer meter relay unit or to a self-balancing voltage bridge unit. If the temperature sensed is below the temperature setting on the pyrometer, a relay device is actuated and heat is supplied to the machine. If the temperature sensed is too high, another relay device turns on the cooling units. Figure 4-2 shows a typical pyrometer controller which can operate in both the heating and cooling modes.

As indicated in the section on machine construction, the usual heating system is an electrical heater. Cooling is usually done by the use of either water cooling or cooling blowers. The call for heat will actuate the heat-delivery unit which can be actuated in one of several different ways. The simplest is a relay unit which will turn the power on until the desired temperature is sensed by the thermocouple and the heat is then turned off by the relay. The cooling units can be similarly operated.

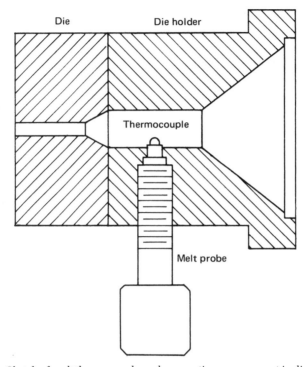

Fig. 4-1. Sketch of melt thermocouple probe mounting arrangement in die holder.

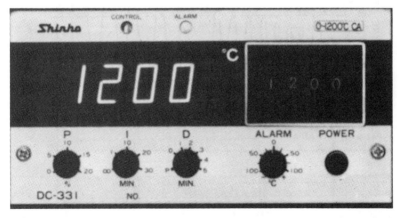

Courtesy Shinko of America
Fig. 4-2. Typical electronic digital temperature control instrument.

When the temperature goes over the set value as indicated by the thermocouple, the cooling is turned on until the set temperature is reached, at which point the cooling unit is turned off. This simplified method of operation suffers from some limitations due to the machine construction and sensor placement.

The thermocouples are placed at some depth in the extruder barrel wall. Since the barrel itself represents a large thermal mass, the heating and cooling units placed on the surface of the barrel result in a lag in the response of the heating and cooling effects. As a result of this, the actual temperature of the barrel rises to a value higher than the set temperature on heating, as the heat transfers from the heating elements to the rest of the barrel as thermal equilibrium is established. This swing in temperatures tends to upset the material temperatures and destabilizes the process. Figure 4-3 shows a time graph of temperature with this condition prevailing.

With respect to this figure and the subject of control, it would be useful to define the terms shown. The set point is the desired temperature. Depending on how accurately the desired temperature is to be maintained and on the characteristics of the controller there is a band for which the control function does not operate. This is called the dead band or dead zone.

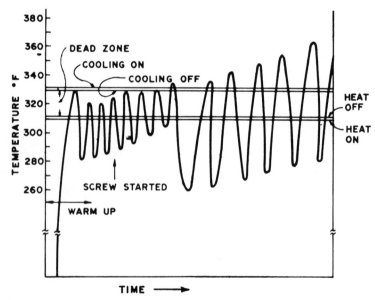

Fig. 4-3. Temperature time variations on the extruder barrel using on–off control.

One of the ways to increase the accuracy of the temperature control is to change the way in which the heating or cooling is applied. Instead of turning the heat full on when heat is called for, the intensity of heat is adjusted to the difference between the sensed temperature and the set point. This type of control is called proportional control. If the sensed temperature is very far from the set temperature, then the heat (or cooling) is turned on high. If the sensed temperature is close to the set point, then the heat or cooling is turned on very slightly. As the sensed temperature approaches the set temperature, the rate of heat input is dropped and is shut off as the temperature approaches the dead zone. The band of temperatures in which this occurs is called the proportional band. A time–temperature diagram of a barrel section controlled in this manner is shown in Fig. 4-4.

There are several ways in which the proportioning of the heat input can be effected. These are controlled by the electronics in the pyrometer controller. One way is by the use of time-proportioning control. In this setup the controller turns the heat on and off several times a minute. The relative length of the off and on time is adjusted by the instrument based on the deviation of the sensed temperature from the set point. Wide deviation will produce long on versus short off, while small deviations will produce short on and long off times. Another scheme involves the

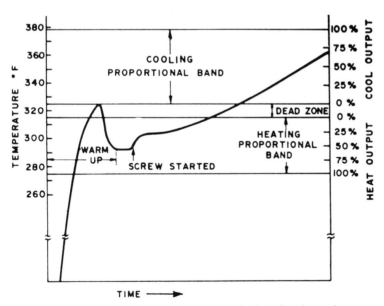

Fig. 4-4. Typical time temperature variations on extruder barrel with simple proportional control mode. Screw: 28:1 L/D, vented; material; PP MI 3.

use of SCRs (silicon-controlled rectifiers) which change the power input to a device by using only part of the cycle of the AC power to deliver power. The portion of the cycle that is active is determined by a control voltage to the SCR units. In the case of the temperature control pyrometers, this voltage is generated by the deviation of the sensed temperatures from the set point. When the deviation is large, the SCR will deliver close to full power. When the deviation is small, the SCR will deliver minimum or no power.

As shown in Fig. 4-4, it is possible for the temperature to still drift to higher than the set-point values. This is a result of having the heat or cooling intensity too high or too low. In most cases the heaters used are substantially larger than the size required for maintaining the temperature since they are also required to preheat the machine on startup. In order to overcome this problem use is made of an automatic reset control function in the pyrometer. This will either reduce the maximum output that can be delivered from the heater (or cooler), or it will shift the point of onset of the reduction in input further away from the set point so that the actual levels of heat input are lower for the same temperature deviation. This function is automatic in the instrument and is brought into operation if the temperature has a tendency either to continuously override the set point or to drop below the set point. The time–temperature diagram for a unit operated in this mode is shown in Fig. 4-5.

It has been mentioned that the extruder barrel represents a large thermal mass and, as such, complicates the temperature control. It has a value, though, in minimizing swings in temperature caused by poor control. The large amount of heat stored in the barrel makes the heat inputs less effective in temperature changes. To enhance this effect when using time-proportioning temperature control, it is desirable to use shorter cycle times for the control. This is graphically illustrated in Fig. 4-6. Changing the time length of a cycle on the controller from 1 min to 0.25 min reduces the temperature excursions by a factor of 2. Another factor in the temperature excursion, caused by the size of the barrel wall, is the location of the temperature sensor. For best sensing of the actual temperature of the inner barrel wall, in proximity to the plastic material, the probe should be deep. For best response to the effects of the heat flows, the sensor should be located near the outer barrel wall. Generally the two requirements are compromised by placing the probes a little more than halfway into the wall. This relies on the thermal inertia of the barrel to act as a stabilizing factor in maintaining the temperature level and gives a better reading of the temperature at the inner wall of the barrel.

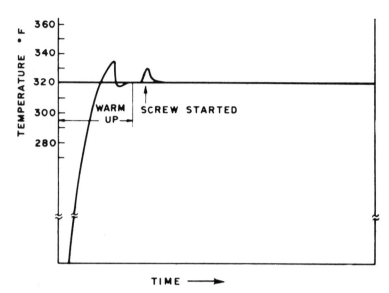

Fig. 4-5. Variation of temperature with time on extruder barrel using proportional temperature control with rate reset feature. Note that the variations in heat input level minimizes or eliminates overshoot.

The temperature controllers used on the dies are heat-only units, since there is no problem with internally generated heat to be dissipated in the material flow through the die. Like the barrel controllers they are usually proportioning instruments, but rarely require the rate reset feature. The die heaters are mainly used to preheat the die and to prevent heat loss from the material as it flows through the die. Occasionally, the die

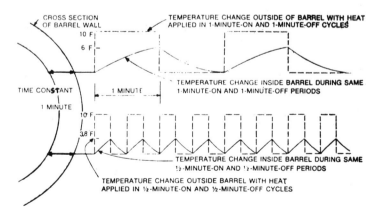

Fig. 4-6. The effect of controller cycle time on the barrel temperature swings using simple on–off control. The thermal inertia of the barrel mass minimizes temperature swings.

temperature must be kept different, usually hotter, than the material flowing through. Then the die temperature control becomes more critical.

The discussion so far has been on conventional discrete instrument temperature control systems. There has been a recent trend to a centralized temperature control scheme in which one instrument is used to control the temperatures for the entire machine. This trend is affected by the tendency to make the extruders operate under the control of a computer, which determines the state of all of the machine conditions and adjusts the appropriate parameters for stable operation. In this scheme a series of sensors are placed on the machine. Their output signals are fed to an instrument that has an adjustable memory system, which stores the settings required for each machine control zone, and a microprocessor, which makes the unit read around each input temperature and compare it with the preset values. If the temperature is low, then the heat level is increased; if it is high, the cooling level is increased for that particular read-around cycle. The cycling time is rapid, several times a minute, so that the temperature is accurately measured and controlled. In most cases, the memory unit has additional information which enables the unit to perform as a proportioning rate reset controller.

One of the more sophisticated versions of the central heat controller also incorporates circuitry which adjusts the barrel temperatures so that the proper melt temperature is achieved. In the discrete instrument system this temperature is usually read, and if it is off, adjustments are made in the barrel temperatures to bring it to the desired range with the understanding that machine speed can also be involved. In the case of the central control unit, data can be stored in the unit which will permit internal adjustment of the set points at the appropriate parts of the machine to return the melt temperature to the desired value. Because these are also affected by machine speed, the data will have to reflect this. In the more sophisticated control systems all of the parameters are controlled to produce the result on line.

One of the other parameters important to the operation of the machine is the material pressure at the extruder head. This is a variable that can be measured directly, but controlled only indirectly. There are several units which vary in complexity employed for pressure measurement. The simplest is a Bourdon-type pressure gage which is filled with a viscous silicone grease. Such a unit is shown in Fig. 4-7 with a valved head used to control pressure. The internal construction of the unit is shown in Fig. 4-8. This particular design suffers from the limitation that the grease can enter the polymer melt and cause contamination.

A different type of unit is shown in Fig. 4-9. This is a grease gage, but it

Courtesy Welex Co.
Fig. 4-7. Melt pressure gage of the open grease-filled tube type.

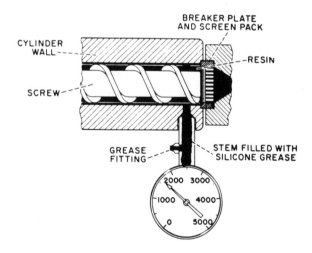

Fig. 4-8. Schematic of grease-filled pressure gage.

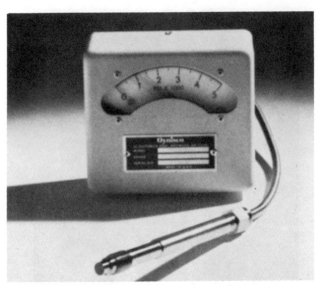

Fig. 4-9. Seal diaphragm melt pressure gage.

has a diaphragm which isolates the grease from the melt. A more elegant unit, shown in Fig. 4-10, has a resistance bridge attached to a diaphragm which measures the pressure. Again, to avoid direct effects of the high temperature on the bridge, the probe used to attach the pressure-sensing unit to the machine is a grease-filled tube which transmits the pressure from the end of the probe to the bridge diaphragm. The output from the sensing unit is converted to a signal which is displayed as a readout on the instrument. It can also be used to operate a control function or to make a continuous record of the pressures.

While they are not widely used, valves are one way in which the pressure can be controlled in an extruder. In many cases the back pressure in a machine is controlled by the use of screen packs which are fitted to the back of a breaker plate. The type, size, and number of screens is arrived at by a trial-and-error process to give the proper pressure drop, so that the machine sees the appropriate back pressure for proper plastication, and that the pressure in the die is appropriate. As the screens catch debris in the melt stream and become partially clogged, the pressure drop increases and it is necessary to change them periodically. Despite the apparent limitation on the technique, it is the one most widely used. Figure 4-11 from Ref. 1 is a graph of pressure drops versus screen construction which can be used to estimate the screen pack requirements.

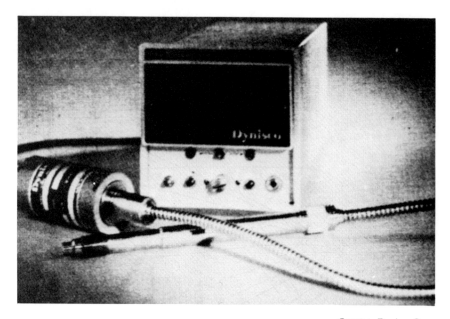

Courtesy Dynisco Corp.

Fig. 4-10. Bridge-type melt pressure transducer with electronic readout.

Valves have become more interesting in recent extruder designs because of the interest in computer control of the machine operation. The valve represents a way of changing the pressure drop and, consequently, the machine back pressure while the machine is operating. When computer-control systems are used they represent a means of feedback control of the machine output. There are various designs for the valves. Figure 4-12 shows the construction of a streamlined valve unit and Fig. 4-13 shows the internal construction of the unit. The obstruction to flow in this unit is the needle valve section on the upstream side. Figure 4-14 shows the internal construction of another type of valve. The flow restriction is between the stationary and adjustable gates where the adjustable gate can be moved from outside of the valve body.

The valves shown are intended for manual adjustment. The designs are changed for the automatic control arrangement by having the valve elements actuated by electrohydraulic servo units, which are capable of handling the melt pressures in the valves, and which can be actuated rapidly in response to electrical signals from the control unit. The pressure in the extruder at the head end is sensed with one of the bridge-type instruments and matched to the desired head pressure setting. If the pressure is different from that set, the valve opens to decrease the pressure and closes to increase the pressure by altering the

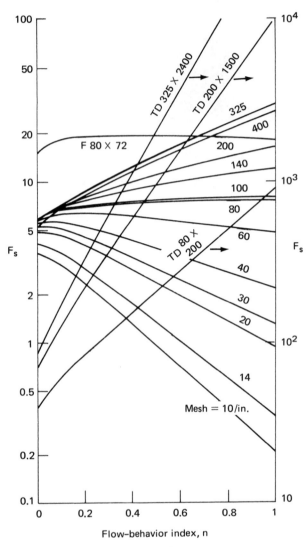

Flow–behavior index, n

Reprinted from Polymer Engineering & Science *18, 409, April (1978).*
Fig. 4-11. Effect of screen pack screens on pressure drop. Screen-resistance factor, F_s, for U.S. standard square woven screens from 10 to 400 mesh, a Fourdrinier woven screen and three twilled dutch woven screens. While the resistance factor generally increases with increasing mesh number (wires/in.), the increase is not steady and interpolation may not be reliable. Some Tyler mesh numbers differ from U.S. standard for the same screen dimensions and must be converted to U.S. before entering this plot. The same is true for metrically designated screens.

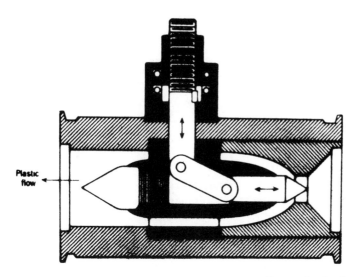

Fig. 4-12. Schematic cross section of streamlined plastics melt valve (Sterling).

Fig. 4-13. Cutaway illustration of a streamlined valve for temperature-sensitive plastics. Adjustment nut at top produces axial movement of nose into seat.

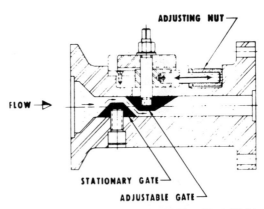

Fig. 4-14. Drawing of another type of streamlined valve. The adjusting nut moves the upper gate horizontally to close the gap (Waldron–Hartig).

gap available for resin flow. As is the case with most adjustments on a single-screw machine, there are other factors which influence the pressure which reacts to the valve adjustment. For example, changing the back pressure will increase the amount of shear work which will reduce the melt viscosity to somewhat counteract the pressure drop caused by the increased restriction and vice versa. In addition, the gap size will generate different amounts of shear on the material flowing through the valve which will also affect the pressure drop. The change in the valve setting to control the back pressure and output is one of the more effective and predictable ways of controlling machine performance, especially when it is compared with the effects of changing the screw speed (another way of affecting the head pressure and delivery rate of the extruder).

Screw-speed control is essential to proper extruder startup and operation. All current production machines are equipped with some type of variable speed-drive system. This is required to adjust the screw speed to the value required for proper melt delivery for the resin used and the screw employed. The speed ranges needed for proper operation vary with the material. For stiff materials which generate large amounts of shear heat and use a deep-flighted screw (for example, rigid PVC and acrylic compounds) the screw-speed range is usually from 5 to 50 rpm. For polyolefin materials using a shallow-flighted screw the speed range is usually 30 to 200 rpm. Other materials such as ABS will use intermediate screw-speed ranges. The range is either preselected in the gearbox for the machine or the gearbox is equipped with change gears to set the range over which the speed-changer drive will operate. There are several types of varispeed units employed, with the specific selection based on cost and the need for greater control.

The simplest type of unit, which was widely used and is still designed into some of the smaller machines, is the variable-pitch pulley unit with block belts typified by the U.S. Varidrive and the Link Belt Drive. The movement of a lever controlled by an external handwheel will open one set of cones and close the other to change the relative pitch diameters of the pulleys and to change the output shaft speed. Although some have been made with electromechanical actuators, this unit is difficult to control remotely. As a result, they have been replaced in the newer machine with more easily controlled units.

One of the earlier changing speed electrical drives used DC motors which had field speed controls. Since DC is relatively unavailable, the systems used either a motor generator set to generate the DC power or a rectifier unit. With the advent of other control modes for the variable speed motors, the DC motors with the field speed control are no longer favored. The complexity and inefficiency of generating the DC power on site, coupled with the nonlinear control patterns of the motors, made them noncompetitive with SCR units and the more recent variable-frequency AC motor drives.

The SCR drives are installed in the bulk of the extruder equipment now being manufactured. They use the cycle clipping of the SCR units to supply varying levels of DC energy as rectified spikes. The amount of power is controlled by the firing voltage of the SCR which is an input low-level signal voltage. This voltage is generated by a special circuit. The level can be set by means of a simple potentiometer control which is remote from the motor and is connected to the SCR control unit by simple wiring. Another advantage of the SCR units is that they can be equipped with a tachometer speed control. This can be operated by the back emf of the motor or by means of a special tachometer generator unit. If the tachometer feedback voltage deviates from the set voltage, an additive or subtractive signal can be added to the signal level which fires the SCR units to adjust the speed back to the set value.

The speed regulation with the SCR motors is very good. Using back emf tachometer signals, the regulation is usually 1% and even under severe load shifts the regulation is under 2%. With separate tachometer generator units the regulation can be held to $\frac{1}{4}$ to $\frac{1}{2}$%. This degree of regulation is usually more than adequate to operate the extruders for constant output. In fact, the tachometer generator system is generally not used, since it is desirable for the extruder speed to shift under load surges, and the counter-emf system does better in controlling stable output.

The SCR motor-speed control lends itself to closed-loop output control of the extruder and seems to be an attractive way to deliver uniform product delivery. Again, there are some complicating factors. The delivery rate from the extruder is not directly related to the screw speed because the single-screw machine is a "leaky" pump. Increasing the screw speed increases the amount of shear on the material and changes the viscosity, which increases the back flow of material. In addition, the screw speed has a direct effect on the location of the melt-bed transition points as described in the section on extruder operation. As a result, the melt bed can shift along the barrel and, in cases where the machine is operating close to its maximum throughput condition, can lead to unmelted material entering the metering zone and causing surging of the output. Unstable surging output is the most serious problem in machine operation, and once the condition starts it is difficult to correct. Frequently, it is necessary to shut down the operation and restart it in order to get stable operation again. As a result of this problem the use of speed control as an output stability rate control is difficult to implement.

Another variable-speed motor drive is used in some of the more recently built machines. This is an electronic device which converts the line frequency of the AC to a variable frequency source used to drive an AC motor. Two types of motors have been used. One is a large squirrel cage conventional AC unit which has speed regulation with load changes of 2-3%. For closer control of the speed to $\frac{1}{4}$% or less, the drive motors are synchronous motors. Again, the superior speed control of the synchronous units are probably counterproductive in extruder operation. Load surges can be damped in the output if the motor does slow slightly when extra resistance is encountered in the machine.

As in the case of the SCR units, the control mode for the variable frequency AC drives are signal voltages generated by potentiometer settings. In these drives the control voltages are used to change the frequency of the control oscillator which produces the AC. In most of these systems the oscillators operate at many times the line frequency and they are counted down to the frequencies actually used. The generated frequencies are used as the control signal to large power transistors which generate the heavy currents required, and the power is generally passed through a transformer to generate the appropriate motor voltages and to isolate the transistor amplifiers from the motor reactions. While tachometer feedback can be employed in the AC system, it is generally not used. Tight speed control is effected by the type of motor selected. Since the speed control is an open-end system, it is less likely to react unfavorably to line transients which can shut down an SCR drive.

The choice of the drive and control system will be dictated by how accurately the speed must be controlled and the cost of both the equipment and the operation. At present the choice is the SCR units but with lower-cost electronics for the AC drives they may be preferred since the motor weight and cost is generally less for an AC unit than for an equivalent DC unit. In addition, the maintenance on AC motors which do not have commutators to wear is usually much lower. In the near future the trend will probably shift to the AC drives as they are made available in larger sizes with lower cost electronics packages.

The last of the control elements in the process to be covered is not the machine but in the downstream equipment. It is apparent that whatever product is made by extrusion it is necessary to remove the product from the die exit under suitably controlled conditions if any dimensional control is to be maintained. This holds true in product which ranges from sheet and film to pipe and profile. Some sort of mechanism must be used which will draw away the product at a constant and controllable rate.

The devices to do this are pullers of one type or another. In the case of the sheet lines the three-roll stack represents a special type of puller unit. In most other cases, a set of pinch rolls or a belt puller is used to move the product through the downstream fixturing as it pulls it away from the die. Several of the puller units are shown in Fig. 4-15. Each of the units has a variable-speed drive with good speed regulation to control the line speed of the product. Assuming stable output from the extruder, the stability of the puller speed will determine the dimensional accuracy of the product.

In most installations the control of the puller is an open system. The motor drives are either the belt Varidrive units or SCR motor drives. The operator will adjust the drive speed until the product is brought into the proper dimensional range. The size is monitored by periodic checking of critical dimensions, and the speed is adjusted to bring the size back into specification if it drifts out of dimension. The size variations can come either from variations in the output of the extruder or from variations in the performance of the puller. These variations are difficult to control and make any closed-loop control difficult.

In addition to the fluctuations in extruder output caused by slight material characteristics changes and changes in the factory environment, the main source of dimensional shift is caused by changes in the line speed of the product caused by the puller. There are several sources of these variations, such as drifts in drive speed, slippage of the product in the pull rolls or belts, or some change in puller characteristics caused by shifting temperatures in the product. The drive speed changes are significant in the mechanical variable speed units. As the belts heat up

Courtesy Foster Allen Co.
Fig. 4-15. Several different pullers used on plastics tubing and profiles.

during operation, there is some additional slip which is erratic and will cause differences in puller speeds. The SCR drives can be made very stable by the use of tachometer generator feedback control, and the regulation of speed can be made to less than ¼%. In this case the major line-speed variation is in slippage.

None of the pulling systems are positive gripping devices. They rely on friction between the product and the pulling element to move the product. As a result, the product will slip to some extent in the pulling machine. The slippage may be very small as in the case of the three-roll stack used on sheet where there is a large-contact area and high-contact pressure. It can be substantial in the case of pipe and tube where even a contoured belt will have a limited contact area and where, in the case of thin-walled parts, the contact pressure is limited by the fragility of the part. In the case of some of the profile sections there is even less traction, and this can lead to severe slippage.

There are two ways of controlling the pulling speed to overcome the slippage problem and other erratic puller performance. One way is to measure the line speed of the product with a rolling contact wheel which makes contact with the product and generates a signal proportional to the line speed. The wheel must have excellent frictional characteristics with respect to the product and be properly contacted to the extrudate or it will slip and give an erroneous signal. The signal can be processed so that it can be used to add or subtract from the control signal for an SCR drive, and in this way regulate the product line speed to a constant value.

The second approach to controlling the puller units to compensate for fluctuations is by a feedback measuring system on critical part dimensions. In this arrangement the variations in the puller and the extruder are both compensated for. This automatic dimensional control by on-line gaging will be discussed in detail in a later section, with a description of the necessary gaging sensors and feedback modes.

Summary

This section has covered the control requirements of the critical extrusion parameters needed for stable operation of an extrusion line with uniform product output. The extruder conditions which can be directly controlled are the screw speed and the barrel and die temperatures. The other parameter which can be influenced is the pressure. This is done by changing the restriction to flow at the end of the barrel with screen packs or with a valve. Most current extruder systems use discrete zone controls for temperature. There is a trend toward central processor control of all zones to produce a desired melt temperature. The use of closed-loop control of the extruder is possible using either valved control of back pressure or control of screw speed. The control of the coacting puller unit is the other element affecting stable product size, and while current practice is to use open-loop control, the trend is to closed-loop control based on sensing either line speed or product dimensions.

Reference

1. J. F. Carley and W. C. Smith, "Design and Operation of Screen Packs," *Polymer Engineering & Science* **18**(5), April (1978).

Chapter Five
Heat-Transfer and Heat-Content Considerations

Simply stated, the extruder is a machine for melting plastics materials and producing a pressurized melt. The remainder of the process (after the melt is squeezed through the die) involves cooling the hot melt extrudate so that it solidifies into the desired shape. Consequently, heat and heat transfer play a major role in the process and must be considered in both the machine design and operation and in the handling of the product.

There are several sources for the energy input to the extrusion process, and a number of ways in which the energy in the form of heat is removed from the process. Figure 5-1 is a diagram of the process with the energy sources and heat sinks displayed. By a thoughtful analysis of this diagram the location of the critical energy control points in the process can be identified and analyzed to determine how they can be controlled for best process operation.

The black areas in Fig. 5-1 indicate thermal energy flows, and the shaded areas indicate mechanical energy flows. In the process all of the mechanical energy is converted to heat by one sort of frictional dissipative process or another. The primary mechanical energy input is provided by the drive motor which converts electrical energy to mechanical power to drive the screw. The drive unit, which includes the gearbox, has some mechanical power converted to heat by friction in the bearing systems, generation of heat by viscous effects in the lubricants in the gearbox, and eddy current and other electrical losses in the motor. The net of these losses represents the mechanical energy introduced into the extrusion process.

The mechanical energy is used to convey the polymer through the machine barrel while it is being plasticized and to build up pressure in the melt. The melt pressure is a form of mechanical energy referred to as the potential energy of fluid pressure. In the process of building the pressure in the melt and conveying the material through the machine the screw shears the polymer and generates heat in the polymer material by viscous action. From the relative size of shaded areas in the diagram it is apparent that the major portion of the shaft work done by the screw is converted to heat in the example in Fig. 5-1. This is generally the case, but it varies with the materials extruded. In the case of material such as rigid PVC which develops large amounts of frictional heat, the proportion of mechanical energy used for viscous shear heating is large as compared with the amount used for conveying and pressure buildup. In the case of materials such as polyethylene with much lower frictional

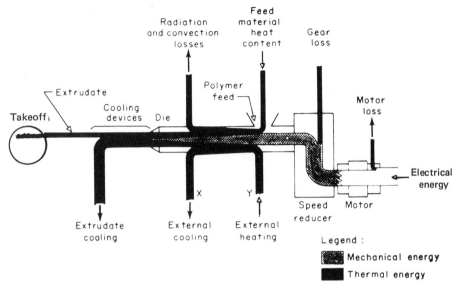

Fig. 5-1. Energy flow diagram for the extrusion process showing inputs of mechanical and heat energy and outputs of heat energy.

characteristics in the melt, a much lower proportion of the shaft power is used to generate viscous heating of the polymer.

The residual energy in the melt, in the form of pressure potential energy, is converted to heat by the pressure drops in the die and the screen packs in the head of the machine. This is a necessary condition since the melt exits from the machine at atmospheric pressure and any pressure energy must be dissipated. The manner is which the final conversion of the mechanical energy to heat is done is important to the performance of the dies.

The heat input into the system comes from two other sources in addition to the conversion of shaft work to heat. One source is the heat in the feed material. The second is heat which is introduced through the extruder barrel walls with the barrel heater system. Heat is removed from the extruder by two mechanisms. The first is radiation and conductive losses to the surroundings, and the second is the controlled cooling system used to maintain the temperature of the barrel wall. Net of these heat flows the material leaves the extruder with all of the heat introduced into the process.

It is apparent that the heat and energy flows in the extruder are interrelated in some fairly complicated fashion. The question of whether heat is added through the barrel walls, or removed, is related to the nature of the material and the process rate. The heat flows are also different during the startup and stabilization part of the process. It would be worthwhile to examine several different cases of materials and rates to understand the direction and magnitude of heat flows.

The first example would be a material with relatively low frictional heat-generating characteristics, such as the polyolefin resins. In the startup phase of the operation the entire heat introduced into the material will come from the heat in the feed stream and the heat transferred through the barrel wall. As the process is adjusted to running conditions with higher screw speeds, a substantial amount of frictional heat is generated in the material and less heat is added to the melt through the barrel wall. If the process rate is increased, the rate of heat generation is increased in the same volume (the extruder contents) and a state can be reached where the total heat requirement is generated by frictional heat. The only function of the barrel heaters would be to supply heat lost by convection and radiation to the surroundings. This condition has been referred to as adiabatic operation even if it does not meet the strict thermodynamic requirements of an adiabatic process. With low frictional heat materials, such as the polyolefins, it is possible to require the barrel to remove heat, but this is rare.

To increase the capacity of the extruder to process the polyolefins, the material fed is sometimes heated to raise its temperature. This heat will minimize the requirement for heat throughput through the extruder barrel walls. Since the material will melt more easily, the load on the motor will also decrease, resulting in lower power requirements for any specific delivery rate. This is especially useful with the low-melt-index resins used to make pipe and profiles because they tend to produce higher mechanical loads on the machines.

In terms of heat transfer, the situation is markedly different when a material with high frictional heat characteristics is processed. As in the other case, the initial startup phase involves transferring heat to the resin from the barrel wall. As the process is stabilized, the frictional heat generated will be sufficient to completely melt the material and there will be no heat introduced from the barrel. In fact, the heat generation usually exceeds that required for melting, and heat must be removed through the barrel walls by the cooling units. This is especially true at higher output rates where the extracted heat is generally as much or more than that required for melting. It is important that the heat be removed since overheating of the resin can lead to unstable flow and to thermal degradation of the resin.

One of the important considerations in the design of equipment to extrude materials with high frictional heating characteristics is that there be adequate cooling capacity in the machine. There are several elements in this; one is the area of the barrel wall. With a given machine size, the only way in which the heat-transfer area can be increased is by the use of longer-barrel machines. A second consideration is the efficiency of the cooling on the barrel. A water-cooling system is usually required for these materials. In addition, the system ought to be capable of handling large amounts of heat. This means the use of large cooling lines and either evaporative cooling effects or strong turbulent flow of the coolant. The cooling units must make good thermal contact with the barrel to avoid air gaps which can cause reduced heat transfer.

Another way to increase the capacity of the machine to remove heat is to cool the screw. The added cooling area, plus the fact that the polymer is moving rapidly past the surfaces of the screw, will usually produce good heat transfer. Since the screw cooling introduces a different set of conditions in the melt bed of the machine it can alter the manner in which the extruder functions. This cooling mode should be used with discretion. In many cases it improves the operation of the machines, but it can severely affect the output in other cases.

In Ref. 1 the interaction of the different elements in the extruder is discussed with emphasis on the effects of the heat transfer on the machine performance. The study indicates one of the sources of unstable flow behavior is due to the heating mode employed. The reference also discusses the effects of the screw configuration on the frictional heat generation and the influence of screw cooling on the extruder performance. Since the melt shear is an important source of heat for melting the plastics material and the magnitude of the heating is dependent on the way the machine is run, heat transfer must be considered along with the other conditions in the machine to predict the extruder performance.

In the case of the high-frictional-heat-generating material, the screw designs are important for the prevention of overheating. In general these machines are equipped with deep-flighted screws to increase the rate of transport of the material at lower screw speeds. This has the twofold effect of reducing the amount of shear and of increasing the effective heat-transfer area on the screw. In most cases the screws are cooled to improve the rate of output.

Most materials extruded fall between the polyolefins and rigid PVC in terms of frictional heat-generating characteristics. As a result, some will operate with heat delivered to the material most of the time and some will operate with cooling most of the time. The heat-transfer requirements are not nearly as severe as those for rigid PVC, but they must be adequate for proper process control.

The next portion of the energy flow to be considered is the die and other exit-end elements in the process. In this section of the machine the pressure energy of the material is converted to heat by pressure drops and straight-line flow through the die. Part of the energy conversion takes place in the screen pack, choke, or valve which is used to maintain a back pressure on the extruder so that proper melting can be achieved. This results in the material undergoing additional shear heating, but at a lower magnitude than the effects in the screw. However, this effect is not negligible and must be taken into account in determining the properties of the extrudate. This is especially true in profile extrusion since the effects will directly affect the melt strength of the material for profile retention.

Another part of the pressure energy is converted to heat in the die itself. This is primarily in the surface layers of the flowing material since the shear rates are highest near the die walls. In a complex die (such as a profile die) this will result in uneven heating. The reason is that the shear rates vary from one section of the die to another and the highest heating effect is in the regions of highest shear. The effect is to soften the melt at these points. Another effect in polymers which are readily heat degraded, such as rigid PVC, is burning and degradation in the high-shear regions.

The general practice is to heat dies so that they will maintain the temperature of the material as it flows through the die. In some instances the dies are operated at temperatures above the melt temperature to decrease the viscosity of the superficial layers of the melt and to reduce the pressure drop in the die. The heated dies can accentuate plug flow, resulting in better shape retention of the die opening and, in the case of profiles, better product dimensional control.

For effective control the dies are made of materials which have fairly high thermal conductivities. One of the problems is that the distance from the die openings to the heaters varies due to the placement and shape of the heating elements and also due to the die shape itself. The differences in conduction path length can result in variations in the temperature at the die wall, which will have significant effects on the extrudate. This is one reason that stainless steels are not the most desirable materials for a complex die shape, since the lower thermal conductivity of these steels will accentuate the temperature differences due to conduction path length.

While there are heating element effects that create flow changes, they are usually not too critical. One reason is that the heat transfer from the die walls to the extrudate stream is limited by low heat-transfer coefficients. In just about every case the plastic material is moving in laminar flow, and this precludes any convection-type heat transfer. Since the thermal conductivity of the polymer materials is low, the net result is that the heat transfer from the wall and through the polymer material is limited.

A detailed analysis of the heating and cooling effects on temperatures is given in Ref. 4. Figure 5-2 shows the manner in which the temperature profile develops in a pipe wall as it is being extruded, and Fig. 5-3 shows the developing temperature field in a slit die. Using the analytical scheme in the reference it is possible to predict the temperature and heat-transfer effects in the die region of the extruder. As in the case with the machine heat situation, the problem is made complex by the heat generation in the materials as a result of the shear.

The shear in the machine also produces a complicating factor in the temperature characteristics of the material leaving the end of the screw and entering the die holder via the breaker plate. As a consequence of the differences in shear history of the material from each section of the screw, the temperature varies across the width of the stream. Figure 5-4 shows the cross-stream variation in a polypropylene material, and it also indicates the variability of the temperatures with time during normal running conditions. Given the sensitivity of the viscosity of the polymers to temperature, these rather large temperature differences can complicate the flow pattern in the extrusion dies. In the cases where this is

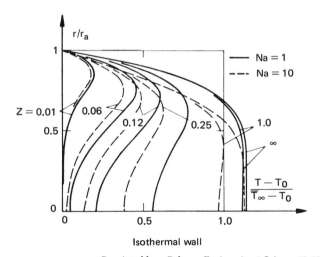

Reprinted from Polymer Engineering & Science **15**, 87, Feb. (1975).
Fig. 5-2. Developing temperature profiles in a pipe extrusion die with isothermal die walls.

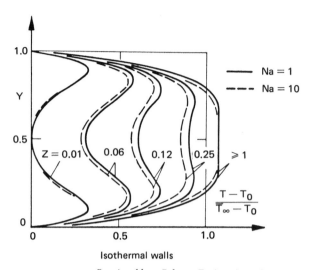

Isothermal walls

Reprinted from Polymer Engineering & Science **15**, *87*, Feb. *(1975)*.
Fig. 5-3. Developing temperature profiles in a slit extrusion die with isothermal die walls.

critical, static mixers such as the Koch mixer shown in Fig. 5-5 are used to mix the streams intimately and to eliminate the temperature variation. These mixers also make excellent heat-exchanger units to produce accurate melt temperatures. They have been used in the production of solvent-blown foams to reduce the temperatures of the melt by as much as 60°F (15.6°C) in a 6D unit.

The heat-exchange and thermal considerations in the takeoff equipment represent an important factor in producing quality product. Heat is removed by a variety of different devices using conduction, convection, some radiation, and evaporative cooling. Each of the heat-transfer modes can be used, and they are frequently used in combination to cool the extrudate properly. One of the most widely used techniques involves conductive heat transfer.

The three-roll stack used in making sheet is an example of a conduction cooling device. The rolls are cooled internally with circulating coolant, usually water, to make the rolls as cool as possible. The hot sheet is contacted intimately with the roll surfaces and heat is conductively transferred from the sheet to the rolls. In this cooling operation the limiting factors involved are the surface area of the rolls and the thermal conductivity of the plastics material in the sheet. Since the contact time of the sheet to the roll is limited with high line speeds, the rolls need to be as large as possible. The trend has been to larger and larger units. Some current machines have rolls 60 in. (1.5 m) in diameter and larger when

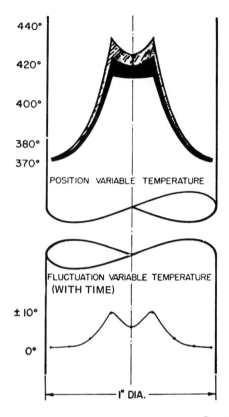

440°

420°

400°

380°
370°

POSITION VARIABLE TEMPERATURE

FLUCTUATION VARIABLE TEMPERATURE
(WITH TIME)

± 10°

0°

|—————— I" DIA. ——————|

Courtesy Koch Engineering Co.
Fig. 5-4. Temperature variation with location across the barrel and variation with time of
the temperatures at the screw tip while extruding 0.3 M. I. polypropylene in a 28:1 L/D
vented extruder. Note the large temperature variation with position.

they are used to extrude thin sheet material. Since the thicker sheet
travels at lower line speeds, the surface cooling of the rolls is adequate.
The low heat conductivity of the plastics material limits the setting by the
rolls. In this particular operation the roll stack sets the surface of the
sheet, and the sheet is further cooled by using either air streams or
cooling water tanks to complete the cooling process. In these units
convection cooling is the mode of heat transfer, and calculations for heat
removal require some knowledge of the heat-transfer coefficients for gas
and water streams.

One condition that occurs in some thick-sheet operations, and carries
through to any thick extrusion, is worth noting. The surface cooling
sometimes produces a rigid layer on the extrudate with a relatively fluid
core. If the heat is transferred out of the sheet too rapidly, the case-

Koch

. Kenics

Courtesy Koch Engineering Co.
Fig. 5-5. Koch and Kenics static melt blenders used to reduce cross stream temperature
variations.

hardened layer remains rigid enough to cause the development of voids
in the extrudate as the material shrinks outward toward the skin. Rigid
materials such as polystyrene and rigid PVC are especially susceptible to
this problem. It requires that the cooling rate at the surface be adjusted to
match the internal conduction rate in the material to avoid voids. In
order to do this, the slower cooling systems are usually used in the
operation after the sheet leaves the three-roll stack. This can be followed
later in the train by a water bath or cooling sprays using water. The
determining factor as to when in the extrudate line to apply the water
cooling is the point at which the sheet or other extrusion is set to a solid
right down to core layer.

Another conduction cooling device which will be discussed in more
detail in the systems section is the vacuum-sizer unit. In some instances,
such as the extrusion of tubing or pipe, the mandrels that form the outer
shape are immersed in a water tank so that there is direct water contact
with the extrudate surface. In this case there is some convection cooling
between the sections of the mandrel and some conduction cooling to the
mandrel sections. In other cases where the dry sizer is used, the sizing

unit is not immersed in coolant, and the extrudate is held tightly against
the forming surface by applying a vacuum to the slots in the sizer blocks.
The heat is removed from the extrudate by conduction to the metal block.
In order to get efficient operation of the block it must be water cooled by
means of cored coolant passages. The coefficients of thermal conductivity of the plastics material and the cooling block materials both affect the
heat-transfer rate as well as the heat-transfer coefficient between the
extrudate and the block.

The heat-transfer coefficient between the extrudate and the cooling
surface will depend upon the intimacy of contact between the two
surfaces. In the case of the three-roll stack, the contact pressure is
determined by the stretching rate and by the nip pressure in the rolls. In
the case of the vacuum sizer blocks, it is determined by the level of
vacuum and the surface character of the plastics material and the block
material. The vacuum level is limited by the amount of drag which can
be tolerated by the extrusion without distortion. With good lubricity
between the surfaces (which can be achieved by using a TFE-
impregnated aluminum oxide coating on the block and a lubricating-type
material such as one of the polyolefins) the vacuum level can be much
higher than when using a chrome-plated brass mandrel and PVC. The
effect of the different levels of contact is to increase or decrease the
thickness of the film of air which is dragged in between the surfaces.
Since air is a relatively poor heat conductor, it represents the major
resistance to the heat flow from the extrudate to the cooling surface. It
also represents a source of variation in the heat-transfer rate since the
film thickness will depend, in a sensitive manner, on the exact operating
conditions during extrusion. Values for the thermal conductivity of air
can be obtained from Ref. 5, and a discussion of heat transfer in plastics
sheets is discussed in Ref. 6.

The other modes of cooling the extrudate involve either air cooling
with fans, blowers, or air jets or the use of water in the form of immersion
baths or sprays. The main mode of heat transfer involved in air cooling is
convection, and the controlling factor is the convection heat-transfer
coefficients and the temperature of the air. In Ref. 7 the mechanisms of
convection heat transfer are discussed and the manner in which the
amount of heat transferred is estimated. The essential parameters to
control are the air velocity, the air temperature, and the heat-transfer
area, which will depend on the length of the extrudate being cooled.

The water tank is a widely used cooling device, and its effectiveness
can be calculated by using an analysis of the convection cooling by the
water in the tank. Reference 7 also discusses heat-transfer conditions
with liquids. In the case of the air cooling, generating high velocities of

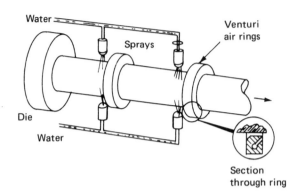

Reprinted from Plastics Machinery & Equipment 6(9), 29, Sep. (1977).©
Fig. 5-6. Combination venturi air cooling and evaporative spray cooling arrangement.

air does not seriously affect the extrudate. However, the case of increasing the water circulation in the cooling tank so that the extrudate has high-speed water jets impinging on it can cause some distortion and deformation of the parts. Also, additional guide units for the extrudate in the tank will be required to prevent such severe displacement of the extrudate that it hits the tank walls. The heat-transfer coefficients in the tank tend to be fairly low and approximate the natural convection values rather than values associated with high-speed turbulent flow.

One of the more efficient cooling systems uses evaporative cooling of the extrudate. The cooling water is sprayed on the extrudate and an air jet is directed onto the material at the same location. The air will cause rapid evaporation of the water from the extrudate surface and will give highly efficient evaporative cooling. From Ref. 7 it can be seen that the coefficients are 10 to 20 times the magnitude that they are for convection cooling. The result is that the cooling train can be substantially shorter than with the tanks. Reference 8 shows the design of downstream cooling equipment in some detail. Figure 5-6 illustrates the arrangement used in a spray evaporative cooling unit described in the reference.

Summary

The heat and heat-transfer relationships of importance in the extrusion process involve heating the incoming material and plasticating it, and, subsequently, cooling the extrudate after it exits from the die. The heating is complicated by the shear heating effects caused by the action of the extruder screw, so that a complete thermal and mechanical heat analysis is required to determine the direction and magnitude of the heat flows involved. The shear heating also complicates the thermal transfers in the dies, but these are much easier to analyze since the material is in viscous flow through the die and the total heat transferred is small compared to the heat content of the material. The other heat-transfer considerations involved are cooling the extrudate. This must be done at a controlled rate to avoid voids, warpage, and other defects in the product. Heat transfer from the product is done by conduction and convection cooling using metal surfaces in the conduction cooling and either air or water as the fluids for the convection cooling. In some cases evaporative cooling can be used. This is very effective in reducing the length of the cooling train especially when large amounts of heat must be removed from heavy sections. The use of standard heat-transfer relationships is effective in estimating the heat removal rates, and this enables the determination of the distances required in the cooling train to cool the product properly.

References

1. Z. Tadmor, "Fundamentals of Plasticating Extrusion—I. A Theoretical Model for Melting," *Polymer Engineering & Science* **6**(3), 185, July (1966).
2. D. I. Marshall and I. Klein, "Fundamentals of Plasticating Extrusion—II. Experiments," *Polymer Engineering & Science* **6**(3), 191, July (1966).
3. I. Klein and D. I. Marshall, "Fundamentals of Plasticating Extrusion—III. Development of a Mathematical Model," *Polymer Engineering & Science* **6**(3), 198, July (1966).
4. J. J. Winter, "Temperature Fields in Extruder Dies with Circular, Annular, or Slit Cross Sections," *Polymer Engineering & Science* **15**(2), 84, Feb. (1975).
5. R. H. Perry and C. H. Chilton, *Chemical Engineer's Handbook*, 5th ed., McGraw-Hill, New York, 1973, Section 3, pp. 243–244.
6. E. C. Bernhardt, *Processing of Thermoplastic Materials*, Reinhold, New York, 1959, pp. 87 et seq.
7. R. H. Perry and C. H. Chilton, *Chemical Engineer's Handbook*, 5th ed., McGraw-Hill, New York, 1973, Section 10.
8. S. Levy, "Cooling Techniques for Profile Extrusion," *Plastics Machinery & Equipment* **6**(9), Sept. (1977).

Chapter Six
Downstream Equipment and Auxiliary Units for Extrusion Lines

In the previous discussion the downstream equipment was mentioned without specifying in detail its function in the extrusion process. In order to intelligently design an extrusion line it is necessary to understand the working mode and limitations of the downstream units and how they interact with the extruder and dies to make a product to specifications.

In order to make the equipment understandable in function, it will be described in relationship to its use for making specific products, even though the units are used in more than one product line. Some of the machines are limited to only one application, but most are multipurpose machines adaptable for several applications.

Sheet Lines

Sheet lines probably have units with the least degree of interchangeability with other product lines. The specific downstream units in the line are

1. The three-roll cooling stack
2. Stripping roller units
3. Auxiliary blower cooling units
4. Gaging heads
5. Cut-off units—saws, shears, or hot wires
6. Sheet stackers

The first two units described are specific to sheet lines and are the primary puller units which remove the sheet line from the machine. The stripper rolls are used to remove the sheet from the roll stack and also to maintain tension holding the sheet to the rolls to ensure good thermal contact with the rolls. They are a simplified puller unit, and in this application the speed is synchronized to the speeds of the rolls in the stack. Figure 6-1 illustrates a simple roll stack unit in place in front of an extruder. The roll stack itself is clearly shown. The pull rollers are at the left side of the illustration feeding the sheet into the cutter, a shear. Figure 6-2 is a view from the cutter side which shows the pull roller and the shear unit cutting the sheet.

The pull units are equipped with cylinders which open and close the rolls on the roll stack and the puller unit. These can be seen on the tops of the stack at the left and right sides and, in this case, the units are large pancake air cylinders. In larger and wider installations hydraulic cylinders are used. In the case of the roll stack, the units push the roll bearing

Fig. 6-1a. Three-roll stack sheet takeoff unit showing die entry section and takeaway roller section.

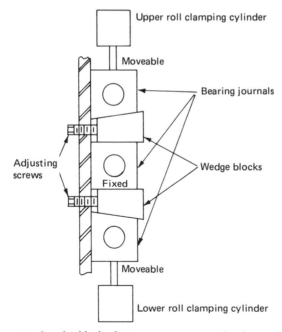

Fig. 6-1b. Schematic of wedge block adjustment arrangement for three-roll sheet stack.

Fig. 6-2. Upstream view of sheet line showing material coming off the three-roll stack along the roller conveyor to the shear.

blocks against a wedge adjustment unit to set the necessary roll clearance. This arrangement is shown in Fig. 6-1a. One of the advantages of the arrangement is that the spacing is held, even though the rolls are opened and closed. The wedge unit gives a precise roll adjustment range since the slope is 1:10; thus a 0.010-in. (0.025-cm) movement of the adjustment screw will change the roll spacing by 0.001 in. (0.0025 cm). The heavy cylinders press against the wedges, holding the rolls closed against the nip pressure. The spacing arrangement is much more positive than any screw-jack unit and is essentially impervious to wear. If the wedge surfaces are kept free of foreign matter, the reproducibility of the gap measurement is less than 0.001 in. (0.025 mm).

In the case of sheet extrusion involving a thicker sheet [0.200 in. (5 mm) and up], the three-roll stack does not completely cool the sheet. The normal practice is to use either fans or centrifugal blowers to direct an airstream at the sheet material to completely cool and set the plastics material. Since the linear movement speed of the sheet is relatively low for thick material, there is usually enough time for cooling in a moderate length and the blowers are effective. For higher line rates involving thinner sheet, the blower units are augmented with sprays of water to improve the cooling rate.

It has become standard practice in most sheet lines to use a gaging head of some type to measure and/or control the sheet thickness. The gages may be simple rolling dial gages at specific locations across the sheet, but can range in sophistication up to scanning nuclear gage units which not only read out the thickness of the sheet, but show the variation in thickness across the width of the sheet and along the length. The signal from such units can be processed by a microcomputer and used to control the die to eliminate variations in sheet thickness. This is discussed in the section on on-line process control.

The units used to cut the continuous sheet to length are saws, shears, and hot-wire cutters. The selection of the appropriate type of cut-off unit depends on the thickness of the sheet and the composition. For thinner gages of most materials a wide power shear is used. The shear can be stationary if the line rate is not too high and the shear operating time is short. If the cut time is long relative to the line rate, the shear is placed on a moving base which is synchronized to the line rate so that the sheet does not buckle.

For thicker materials a traveling saw is the preferred cutting unit. The saw is essentially a panel saw which can traverse the width of the sheet. It is placed on a movable base unit that is synchronized to the line speed so that the movement of the sheet will not cause curved cuts or binding of the saw blade. The cut cycle on this unit, as well as on the shear, is triggered by a limit switch which is tripped mechanically or optically by the leading edge of the sheet. The signal makes the saw go through a tracked cutting cycle and then return to the starting position for the next cut. The position of the limit switch will determine the length of the cut piece.

The shear and the saw generally do not have any significant limitation placed on their use by the characteristics of the materials. Some sheet materials are so brittle that the shear may shatter them when they are completely cold (polystyrene, for example). The alternatives are to switch from a shear to a saw, or to let the material retain more heat so that it is still ductile at the time it is sheared. Some materials, such as high-impact polystyrene and ABS, have a tendency to gum the saw blades, especially if the material is not completely cold and the saw blade is rotating at high speeds. This problem is handled by better sheet temperature control, special low-speed saws, and, if these are not effective, the use of band saws in place of the circular saws generally used.

The hot-wire technique is effective in giving smooth edges to sheet materials where it can be used. Plastics, such as rigid PVC, which decompose to black materials are not cut with wires. On the other hand, acrylic materials, which cut clean and actually form heat-polished edges, are very suitable materials for hot-wire cutting. The hot wire is traversed across the sheet and can be tracked either on a diagonal to compensate for the sheet travel or on a traversing device such as the one used for the saw.

The stacker units catch the sheets as they are cut with a conveyor unit of some type and drop them on a pile which is then packed on a skid. The commercial stacking units used for other materials are adapted for use with the plastics sheets.

Film Lines

As previously mentioned, two different types of plastics film are made, and the lines required to make them involve some differences in the downstream equipment. The first to be covered is blown film, and the equipment required is

1. Venturi ring support
2. Bubble cooling units
3. Cooling and blow-up tower
4. Pinch draw rollers
5. Bubble collapsing slats
6. Winder units
7. Film treater

The venturi cooling rings serve two functions. They are used to stabilize the blown bubble of material by venturi action of air passed outside of the expanding plastic tube. This creates a negative pressure on the material and helps to expand the bubble. Since the effect is a function of the position of the ring with respect to the bubble, displacements of the bubble are minimized. The other function of the rings is to cool the outside of the bubble with the flow of cool air. Figure 3-20 shows the cross-section through a ring with the air passages shown. Figure 6-3 is a photograph of a unit held in position above a film die. It is apparent from the photograph that the air used is low-pressure air supplied from centrifugal blowers. To be effective the venturi action requires large volumes of low-pressure air.

Recent practice has tended toward adding internal cooling to the bubble. This was mentioned briefly in the tooling chapter, and a unit employing internal cooling is shown in Fig. 6-4. The unit shown has several levels into which the cooling air is injected. The use of these units will double the output of a system if the limiting factor in the production

Courtesy Egan Company

Fig. 6-3. Small blown-film die on die cart.

is the rate of cooling of the material (as it frequently is) in high-output systems.

The tower used for handling the bubble is a vertical stand unit which carries the blown-film bubble up to the pinch-draw rollers. A view up the tower from the die is shown in Fig. 6-5, and the collapsing slats can be seen in operation. Figure 6-6 is a side view of a different type of tower unit which shows some of the blown-film bubble guides employed to center the bubble. Each film unit has a tower of this type built to the requirements of the system.

Figure 6-7 is a side view of one type of slat unit used to collapse the blown-film bubble so that it can be brought into the pinch draw rollers. The slat surfaces are covered with a silicone rubber material to minimize the possibility of leaving marks on the film. The slat units have an adjustment to control the collapsing effects so that they have a minimum interference with the bubble.

Various types of pinch draw rollers are used, but typically, the units

Fig. 6-4. Blown-film bubble going into cooling tower. Bubble supports are shown and the bubble is internally cooled.

consist of one polished steel roller and one rubber-surfaced roller driven by a controlled-speed drive to the speed required for the production rate. Figure 6-8 is a photograph of a draw roller unit showing the two rollers with the film over the polished metal roller. The rollers can be opened and closed by means of the air-cylinder drive seen in the illustration. This is needed in the threadup operation so that the tubing can be pulled through to start the run. In some instances the rolls may be internally cooled if there is enough residual heat in the film to heat the rolls. This condition is possible if thicker film is run at very high rates that outstrip

Fig. 6-5. Upper portion of blown-film cooling tower.

the capacity of the air ring and internal cooling units to remove heat from the material.

In most blown-film operations the product is collected by winding onto rolls. Figure 6-9 is a photograph of a turret winder of the type used. The unit is shown winding up the tube material onto a cardboard core in the foreground. The arm carrying the cardboard tube extends back, and there is another cardboard tube in position on the other side of the turret. When one roll is completely wound, the turret flips over and positions the other cardboard tube for winding. Many of the winders are self-threading (as is this one), so that after the film is cut and the turret

Fig. 6-6. Side view of bubble entering the cooling tower.

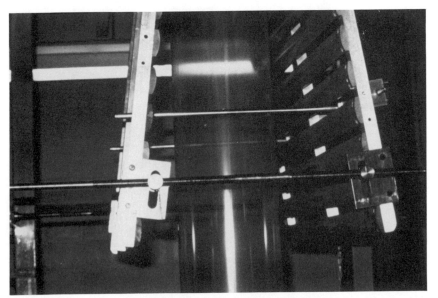

Fig. 6-7. Blown-film bubble undergoing collapse by collapsing slats in cooling tower.

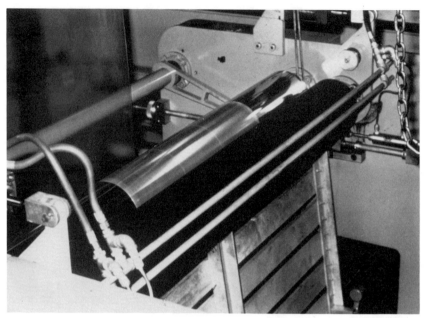

Fig. 6-8. View at the top of the blown-film tower showing the collapsed bubble going over the collecting roll.

operates, the film automatically starts to wind on the new core while the full roll is removed. In some equipment the tubular collapsed film is slit to remove the edges so that two single-layer films are made. Figure 6-10 shows such an in-line slitting stage in a winder where rotary slitter knives are used to trim the edges.

Since one of the major applications for plastics film is in packaging, printing is an important product requirement. Polyethylene and the other polyolefins are among the most extensively used films for this purpose, and they have waxy surfaces that make printing difficult or impossible. To overcome this problem the surface of the film is treated to oxidize the surface layer. The most widely used device to do this is a corona discharge unit. This ionizes the air in proximity to the film surface and the excited ions of oxygen rapidly oxidize the surface of the film. A treater unit to do this operation is shown in Fig. 6-11. The side hoses are used to exhaust the ionized air to the outside since it can be a health problem if it is released into the plant atmosphere.

There are a number of variations of the film process used to make special products. They use embossing operations, laminating operations, and on-line bag-making operations which require other pieces of equip-

Reprinted from Plastics Design & Processing, Feb. (1980).
Fig. 6-9. Turret winder unit winding film after extrusion by blown-film process.

ment. Some of these are mentioned briefly in the section on systems, but the variety is so large that it is difficult to cover all of the equipment.

The other process for film manufacture is the slot-cast film. The equipment differs substantially from the blown film in some respects, but there are units common to the two techniques. The equipment for the cast film includes

1. Casting roll
2. Pinch draw roll unit
3. Winder
4. Treater

Fig. 6-10. Edge trim slitter to making blown film into single layers.

Fig. 6-11. Corona treating unit to make blown film printable.

The special item in the cast-film unit is the casting roll, which is a large-diameter polished roll that is water cooled. Most of the heat in the film material is removed by the chill casting onto the roll. The roll is supplied with a high efficiency cooling system which usually circulates either cold water or antifreeze solutions at high velocity through passages in the roll near the roll surfaces. The pinch draw roll is usually integral with the casting-roll unit. Figure 6-12 shows the film-casting line arrangement with the chill-casting roll and the other equipment such as the pinch draw roll.

The other units in the cast-film line are identical to the ones used in the blown-film line. The film is wound onto a roll in a turret winder unit similar to those used for the blown film. Edge slitting is frequently used to bring the film to a standard width. If the film is a polyethylene or polyolefin material, the same type of ionization unit is used to treat the film surface to make it receptive to inks for printing. Probably the major differences in the line are the subtle changes that must be made because of the different handling characteristics of the blown film and the cast film. The blown film is oriented and generally fairly crystalline because of the slow cooling done in the process. The cast film is as amorphous (noncrystalline) as it is possible to make commercially. This film has much lower stiffness and tensile strength, and requires that the handling equipment apply a minimum of line tension or the film will stretch and distort.

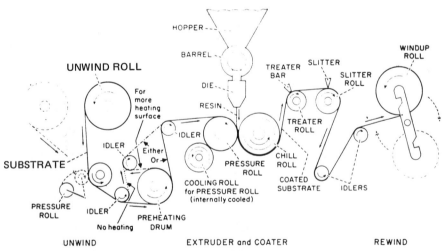

Fig. 6-12. Line diagram for cast film coating of paperboard.

Pipe and Tube Lines

The equipment requirements for downstream in the production of tubular products vary with the size of the product and with the material extruded. The equipment consists of

1. Sizing unit
2. Cooling tanks
3. Auxiliary cooling units
4. Puller
5. Cut-off unit (optional)
6. Winder (optional)
7. Stacker or dump table (optional)
8. Marking unit

Two different methods for sizing tubular products are used when the product has the dimensions controlled by something besides the die opening and the pull rate. These are the internal-sizing cooling mandrel or an external vacuum sizing device. Some materials are extruded into tubing without any secondary sizing unit. This may be because the dimension and shape control resulting from free extrusion is adequate for the product requirement. The other reason is related to the properties of the material. Soft materials, such as flexible polyvinyl chloride compounds or urethane elastomers, are just about impossible to size mechanically because the hot polymer sticks to the sizer surfaces and prevents their movement through a sizer unit. Materials like this are free extruded into a water bath, and various types of fixtures are used to support the tubing and to maintain the shape as the material is cooled. The units are usually rings or rollers and, in some cases, are tubular, loosely fitting mandrels into which water is pumped to produce a high-velocity water flow on the surface of the tubing to improve the cooling rate. In general, the accuracy of the tube diameter and wall thickness is not as good with the free extrusion as it is with the controlled mandrel-type cooling and sizing devices.

The water-cooled internal mandrel shown connected to the die assembly in Fig. 3-29 is designed to size the internal dimension of a tubular product. The taper on the unit must be set for a particular extrusion range. If the extrusion is too fast, the tendency would be to jam the material on the sleeve and stop the process. If the rate is too low, then the drawdown effect will tighten the tube around the sleeve and stop the process. There is some flexibility in the wall thickness that can be made using the internally cooled mandrel, but it is limited.

The vacuum sizing technique is the one which is most widely used to size tubular products, especially thicker wall materials such as plastics pipe. The vacuum tank with internal cooling bushings is illustrated in Fig. 6-13. The tanks have two or more sections with the vacuum applied to the first section. The remaining sections can have either a reduced vacuum level or atmospheric pressure, and are used to complete the cooling of the pipe section. The details of the bushings used in the tank vacuum sizer are shown in Fig. 6-14. Another type of vacuum sizer is shown in Fig. 6-15. In these units the mandrel is cooled by the use of a water jacket or cored holes, and the tube passes through the interior dry portion of the mandrel. It is important that the mandrel have a minimum of friction against the tubing to prevent drag on the extrusion. This can be done by plating the surfaces of the metal in the sizer mandrel with chrome or by using a TFE-resin-impregnated aluminum-oxide coating on a mandrel made of aluminum. A unit of this type is shown operating in the photograph in Fig. 3-37 where the neckdown of the entering tube to make the seal can be seen clearly. Materials with good surface lubricity are necessary to use this type of sizer successfully. In addition, the device works better with thicker wall products which are more self-supporting and have longer residence times in the sizer units because of lower line speeds.

In the case of heavy wall pipe, the vacuum sizing unit does only a part

Fig. 6-13. Vacuum sizer tank for sizing pipe and tubing.

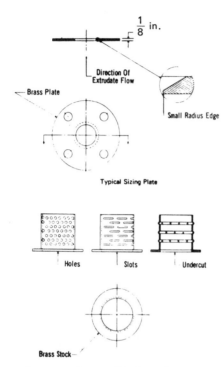

Fig. 6-14. Three typical types of bushings for vacuum sizer tank.

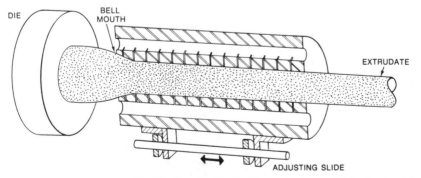

Reprinted from Plastics Machinery & Equipment *6(9)*, 30, Sep. (1977).©
Fig. 6-15. Schematic of dry sizing unit for pipe, tubing, or foam rod.

of the cooling required. The surface of the extrudate is set but the interior
is still soft. To complete the cooling the pipe is carried submerged
through additional water tanks. An alternative cooling system involves
the use of water sprays with or without air applied to effect evaporative
cooling. The venturi air rings described in the chapter on heat and
heat-transfer effects can also be employed to do the secondary cooling. It
is important that the cooling start as soon as the extrudate comes out of
the vacuum sizer so that heat conducted from the interior of the pipe wall
does not soften the set surface and cause distortion or loss of the
controlled dimension.

The next unit in the downstream train is the puller. This can take
several forms, but the unit in most common use is a two-track belt unit
with opposed belts. A unit of this type is shown in operation in Fig. 6-16.
The blocks used are frequently corrugated to obtain additional traction
on the round shape. The units are equipped with variable speed drives to
control the belt speed, and thereby control either the overall tube size in
the case of free extrusion or the wall thickness to a limited degree using
the vacuum sizer or internal mandrel sizing system. As was previously
discussed in the section on puller speed control, the speed must be

Fig. 6-16. Twin track puller unit with gripper blocks for complex profiles.

accurately regulated, and the belt slippage allowed for, to effect accurate and consistent size control.

The smaller sizes of the rigid materials and most sizes of flexible materials can be wound on spools for handling. A spooling unit used for relatively small-size tubing is shown in Fig. 6-17. There are two spools on the unit, so that when one is filled the extrusion can be transferred to the other spool while the first spool is dismounted from the machine and replaced with an empty one. Spooling equipment is available to handle flexible sections that are as large as 3 in. (7.62 cm) in diameter, but the basic design is the same as that of the spooler shown. In some cases the spools are surface driven rather than shaft driven to make control of the line tension easier. All of the spoolers are equipped with a speed control so that the winding rate can be adjusted to take the product at a constant line speed.

The alternative handling system for tubular products, especially rigid materials, is to cut the material into predetermined lengths which are then bundled for handling. There are two ways to cut the product—a flying knife cut-off unit or a traveling saw. The choice is determined by the wall section and by the amount of force that might be required for a knife cut. Another determining element is the quality of the cut. Saws under proper operating conditions will give a squarer and cleaner cut than the flying knife. Figure 6-18 is a photograph of a flying knife cut-off unit with the guards removed to show how the knife is actuated. The usual operation of these devices does not require synchronized movement of the unit with the extrudate line. The cut-off time is so small that

Courtesy Progressive Machine Co.
Fig. 6-17. Winder unit for small diameter flexible tubing or small flexible profiles.

Fig. 6-18. Small flying knife cutoff unit with guards removed to show knife.

the line will not be interrupted by the knife enough to cause buckling of the extrudate. The unit is tripped by a limit switch at the desired length.

A saw is used for the heavier and harder-to-cut sections. As in the case of the heavy sheet materials, it is essential that the saw be on a movable base which can be synchronized to the extrudate to prevent binding the saw blade as it cuts. The saws can either traverse across the pipe, or can cut on a pivot or other arrangement. There is even a saw unit which cuts around the pipe for very-large-diameter heavy wall sections. One saw unit of the swing arm type is illustrated in Fig. 6-19 cutting through a medium-weight pipe section. The swing arm units have the advantage of relatively low drag on the equipment during the cut traverse and they tend to bind less than the traversing saws. Figure 6-20 illustrates the limit switch which is used with this saw. If the direct hit mechanical limit switch is not desired for reasons of safety and accessibility, the usual alternative is a retroreflective photoelectric sensor. The units are so sensitive that cut lengths can be controlled to ±⅛ in. (3.18 mm) and closer if necessary.

The material that is cut off is collected in a unit called a dump table. This is a hopper arrangement into which the cut units fall after being parted from the extrudate. When a sufficient number are collected, they are either tied and tipped out of the hopper unit or are dumped into a box, hence the name dump table. These units are relatively simple in construction.

Frequently, pipe and tubing are marked on line with identification indicia such as the size, wall thickness, or service for the product. This is done with a rolling wheel printer such as the unit shown in Fig. 6-21. The unit rests on the extrudate and is usually driven by the friction between the product and a friction wheel attached to the print wheel. Other units

Fig. 6-19. Automatic tracking cutoff saw.

Fig. 6-20. Limit switch tripping unit to initiate saw cut at predetermined length.

Fig. 6-21. On-line drum printer for marking extruded products.

may be synchronously driven with a motor drive if the friction unit causes problems. The indicia are generally simple so that the print quality requirements are not too severe. As a result, the friction-drive units are generally favored because of the lower cost and simpler operation.

Rod and Profiles

One of the differences between the tubing lines and the rod and profiles is that the shapes can be everything but round. Many of the units that are used with the tube-and-pipe extrusion are used for the rod-and-profile lines. In some cases the units are modified to handle the odd shapes of the profiles. The line usually consists of the following equipment:

1. Fixturing rack to hold part shape
2. Vacuum sizer (optional)
3. Cooling unit—either air or water
4. Puller
5. Cut-off unit
6. Winder (optional)
7. Printer or marking unit (optional)
8. Embosser (optional)

Since the products involved in this category are diverse, the lines used in their manufacture will vary a great deal from each other. The simplest lines will use the minimum amount of equipment with the puller, cooling unit, and a cut-off or winder as the essential items. As the requirements for the product are made more exacting and complex, other units such as shaping fixtures, vacuum sizers, and printers or embossers are added.

A simple extrusion made from a readily profilable material, such as PVC, would be carried from the die to a simple set of fixtures which support the product as it is air cooled and carried away by a two-belt puller, such as the one described in the previous section. Since most of the PVC extrusions are rigid, the material would be cut to length with a saw or flying cut-off unit and stacked. A flexible PVC extrusion would normally be immersed in a cooling tank or trough for cooling as soon as it left the die and then led to a two-belt puller for removal at the controlled rate to size the product. Here the product would probably be rolled on a winder since the flexible materials can be readily coiled.

Many profiles require additional equipment. In most cases maintaining the profile requires fixtures to hold the part in shape while it sets. Some of the different ways in which product is handled to set the material will be discussed. Figure 6-22 shows the construction of a simple water tank. In the illustration the product is led into the tank through a weir which has approximately the shape of the product. This is the preferred method. In many cases, however, the material is led into the tank and removed over the top lip of the tank. The essential requirements of the tank are the leveling units to bring the tank to the correct height for the extrudate line, the water entry for cool water, and the overflow drain pans on the end. Generally the water is not agitated to increase the heat transfer. If additional cooling effects are required, some of the devices shown in Fig. 6-23, such as the cooled guides or the

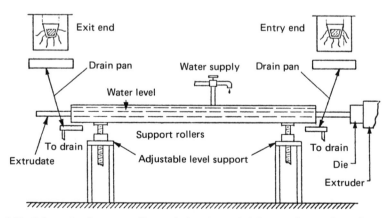

Fig. 6-22. Schematic of water-cooling tank showing weir inlet openings and overflow pans. Height of the unit is adjusted with screw jacks.

cooling blocks, are used. These units serve the dual purpose of supporting the product and increasing the rate of cooling by flowing the water at
high velocity past the surface of the product.

In some cases more sophisticated shaping fixtures are required. The
roll units shown in Fig. 6-23 are postdie shapers for difficult shapes. The
draw plates are used with some materials if the shape has a tendency to
fold over. These units are usually used in conjunction with the water
tank as auxiliary holding fixtures and cooling units. In some cases where
the product cannot take the rapid cooling induced by putting the product
directly into the water, air cooling is substituted.

Figures 6-24 and 6-25 show fixture arrangements used on two air-
cooled profiles. In both cases it is essential that the product be supported
while it sets to prevent the edges of the part in the first case, and the lip in
the center in the second case, from touching the adjacent wall to prevent
welding the two together. The fixtures hold the part in shape as it is set by
the air streams. After the initial set of the part, it can be water cooled to
complete the cooling to room temperature.

Where the dimensions must be closely held, the use of vacuum sizer
blocks and mandrels is essential. Figure 6-26 is an illustration of one

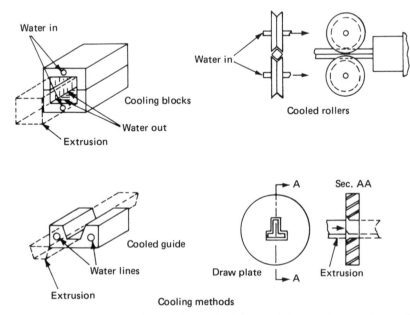

Fig. 6-23. Schematic representation of several cooling and forming fixtures for profile
extrusions. Cooling blocks and cooling guides are usually used in the tanks. The draw plates
and rollers are used above the water line.

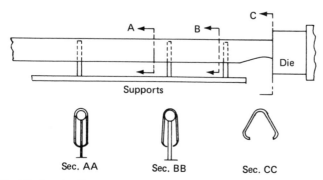

Fig. 6-24. Post die forming fixture arrangement used to make a profile with a small opening slot.

vacuum sizer unit used to size the hollow profile with the attached flat section. By drawing a vacuum on the part the hollow section is accurately sized, as is the flat attachment. In addition, the relationship between the parts is maintained. Because it is essential in this type of sizing system that the surfaces held cause no distortion of the part, only the far side of the flat is held.

Another method of using vacuum sizing techniques is illustrated in Fig. 6-27. This unit is a one-sided vacuum sizer unit where the part is held to the sizer by vacuum to set the shape. This approach is widely used to maintain shapes on moldings and is used to shape vinyl siding materials during extrusion. There are limitations imposed on the use of sizers since the shrinkage of the material will tend to pull the parts away from the mandrels. The limitations on the shapes are indicated in Fig. 6-28. This shows how the pull tendency can be counteracted in some cases by splitting the shaping mandrels and which types of shapes will not work.

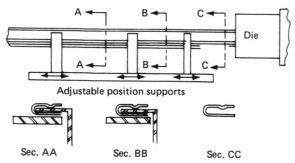

Fig. 6-25. Post die holding and support system for air-cooled profile. The supports are moveable to permit location at optimum points.

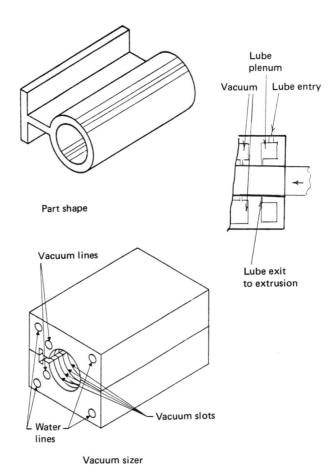

Part shape

Lube
plenum
Vacuum   Lube entry

Lube exit
to extrusion

Vacuum lines

Vacuum slots

Water
lines

Vacuum sizer

Fig. 6-26. Dry vacuum sizer for gasket shape. Note the vacuum ports are located on one side of the T section and the inlet lubricator.

The puller units used for the profiles are the same construction as those used for the pipe-and-tube extrusion. One of the additional requirements in the case of lines that use vacuum sizer blocks is that the power capabilities be on the high side, since the drag produced by the one-sided blocks is generally higher than that existing in a pipe line. The belts are also made to permit good holding on odd-shaped parts. The belt-covering materials should be selected for high friction and should generally be soft so that they will conform to the part shape more easily.

The cut-off units used with profiles are also very similar to those used for pipe-and-tube extrusions. The saw types are generally limited to traversing units for wider extrusions while the pivot units are used for

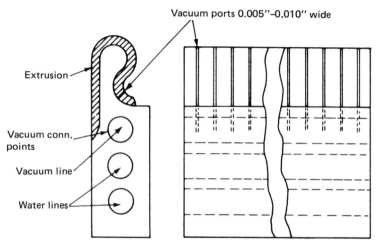

Fig. 6-27. One-sided vacuum sizer for hairpin-shaped part. Vacuum ports are covered by the extrusion for sealing the part to the sizer.

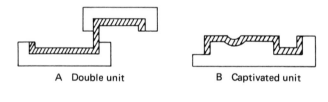

A   Double unit   B   Captivated unit

Sizer sections

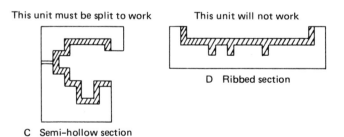

This unit must be split to work   This unit will not work

D   Ribbed section

C   Semi-hollow section

Fig. 6-28. Schematic of several one-sided sizer units showing which are operational in production. The nonfunctioning units will either jam the extrusion or peel the extrusion from the blocks.

narrow sections. The flying knife units are equipped with guide bushings that approximately match the shape of the part to prevent fracture or tears in the parts when they are cut.

Embossing is an operation widely used to decorate such plastics profiles as siding and trim sections. One of the problems encountered in the embossing is that the pressure of the embossing roller will cause uneven cooling and possible warpage of thin wide sections. Figure 6-29 shows two cooling arrangements that can prevent this from occurring. In one case the sizer plate is used to cool the back of the part, and in the other air streams are used for back-side cooling.

The components of a complete extrusion line for profiles can include additional units which will cut shapes on line and reform the parts on line. The units described represent the ones that are most frequently used. The functions involved are holding and cooling, pulling and cut-off, or coil to make the operation complete.

Wire Covering and Coating

One of the distinctive features of the wire-covering and coating operations for small sections is that a material is fed through the die. Usually the operation is a crosshead arrangement with the line disposed at right angles to the extruder barrel. To conserve plant space some of the operations use an offset head so that the operation is laid out parallel to the extruder but offset to one side so that the feed line can bypass the machine.

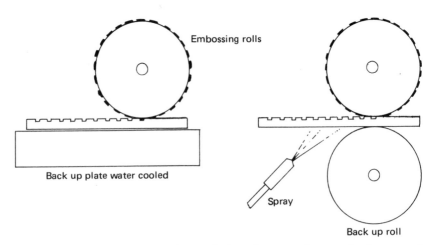

Fig. 6-29. Schematic representation of two techniques for preventing extrudate bow caused by embossing by compensating cooling.

Two distinctly different equipment arrangements are used for wire and other continuous strand covering, which is relatively flexible, and the covering of rigid substrates such as wood dowels or metal shapes. The wire- or strand-covering operation will be discussed first. The elements in the line are

1. Payoff unit
2. Capstan
3. Preheater
4. Cooling unit
5. Second capstan
6. Spark tester (for electrical wire)
7. Puller
8. Winder

The general arrangement of a typical wire-covering line is shown in Fig. 6-30. The units shown are the payoff stand, the first capstan, a preheater, the die, a cooling tank, the spark tester, second capstan, and

Fig. 6-30. Treatment line for flexible core materials.

winding unit. The payoff unit is the web supply and is generally a two-roll unit. This way, when one runs out the second feed roll can be attached to the trailing section of the first roll to make the process continuous. The two capstans are tied together with a synchronizing drive arrangement so that the wire is kept under constant tension and taut as it is fed through the die. To prevent the web material from cooling the plastic melt prematurely it is run through a preheater to bring the material up to the melt temperature of the plastic. After the covered web leaves the die it is run into a water-cooling tank to set the plastic material. In the case of insulating wire, it is desirable to detect the existence of voids or weak spots in the insulation jacket as soon as the wire is made in order to prevent the generation of large amounts of scrap. To do this, the covered wire is stripped of water and run through a spark tester where a high voltage is placed between the jacket surface and the wire. If the insulation is weak, there is a breakdown which is observed and recorded. The line shown does not have a puller unit because it is not required for lighter gage materials. For heavier wire the puller is placed between the second capstan and the winder unit to maintain tension on the line and to make the capstan action effective.

With the exception of the water tank and the puller, the units in a wire or small flexible web-coating line are specifically designed for this operation. In the case of large cable, some of the winding and capstan units have to be made with large-diameter winding capabilities and require large drives to supply the necessary pulling power for the heavy cables.

The covering operation for rigid material such as dowels for towel bars uses a different set of units. The equipment required is:

1. Magazine feed
2. Roll feed and roll former for sections formed in line
3. Bar or dowel feeder
4. Preheater
5. Puller
6. Cooling unit
7. Cut-off unit
8. Finished parts collector

A line for the discrete bar or dowel coating is shown in Fig. 6-31. The parts are stored in a magazine feeder that drops them onto the rollers which carry them into the process. The first of the rollers operates at slightly higher than line speed, with some slippage so that the bars or dowels are butted against each other entering the extruder crosshead. This prevents material from getting on the ends of the bars. The next unit

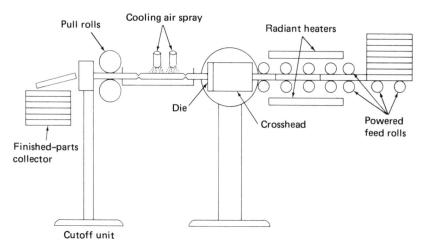

Reprinted from Plastics Machinery & Equipment 7(2), 43, Feb. (1978).©
Fig. 6-31. Schematic of extrusion line used to cover towel bars with plastic.

in line after the feeder is a preheat oven which is shown in the illustration as a radiant heating unit. Forced hot-air units are also used for this step. The requirements are to remove surface moisture from the parts and to bring them up to melt temperature. This permits good adhesion between the coating and the substrate. The puller, which is shown as a set of pull rollers, moves the coated units away at slightly higher speed than that at which the machine is delivering the parts. This results in a separation between the units, making it easy to detect the ends of the units for accurate cut-off. The cooling system shown is a combination of air jets and sprays which is usually very effective in this type of operation. The collecting unit can be either a special stacker or a dump-table unit similar to those used for regular profile and pipe operations.

The other rigid line is used to coat metal shapes which are roll-formed continuously from strip on line. Sections for storm windows have been made this way. Figure 6-32 illustrates a line used for this operation. With two exceptions the arrangement is the same as for the towel-bar coating line. One is the feed-and-roll-form unit which shapes the part that enters the extruder. The other unit that is different is the cut-off device which must be capable of cutting through the metal substrate. Shear-type cutters normally used for cutting roll-form shapes are needed, rather than the plastics-type cutters, and since the line is very stiff, they must be moved in synchronization with the line movement while cutting to avoid cobbles in the line. The line illustrated has a two-belt puller rather than

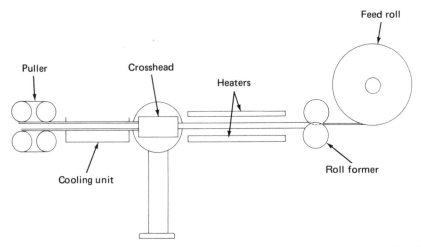

Reprinted from Plastics Machinery & Equipment 7(2), 40, Feb. (1978).©
Fig. 6-32. Schematic of extrusion line to cover roll-formed metal sections with plastics. The output of the roll former is fed directly to the extruder crosshead.

pull rollers, because the pulling force required for the continuous line is substantially greater than that for the dowel-covering operation.

Specialty Lines

Specialty lines usually involve systems which do other operations on line such as forming of the parts and cutting operations on the line. Some specialty line units are shown in the following illustrations from Ref. 1. Figure 6-33 shows a two-belt unit for corrugating a wide, flat extrusion.

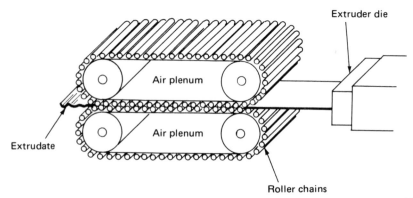

Reprinted from Plastics Machinery & Equipment 7(8), 27, August (1978).©
Fig. 6-33. Twin-belt corrugator unit to make corrugated plastic strip.

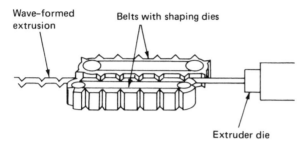

Wave–formed extrusion

Belts with shaping dies

Extruder die

Reprinted from Plastics Machinery & Equipment 7(8), 27, August (1978).©
Fig. 6-34. Postforming extruded plastics rod and profiles by means of a double-belt unit.

The material is cooled by the use of air plenums which are inside the belt structures. Figure 6-34 shows a side-forming unit which produces wave-form extrusions. In this type of unit the belts are generally water cooled by sprays on the return travel and the cooled belt will set the extrudate.

Figure 6-35 is typical of the on-line secondary forming that can be done with extrusions. The extrudate is wound into a tight coil as it comes off the line. This is used for some types of retractable wire, retractable tubing, and special extrusions to make corrugated tubular sections. Large diameter tubing is frequently corrugated by a multiple-belt unit which works in a manner analogous to Figs. 6-33 and 6-34 where the belts surround the tube which is inflated against the belts.

Figure 6-36 shows a synchronized punch which can be used to make cutouts in an extrusion. Figure 6-37 is a rotary die cutter which can be

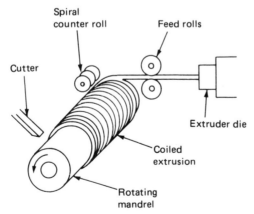

Spiral counter roll

Feed rolls

Cutter

Extruder die

Coiled extrusion

Rotating mandrel

Reprinted from Plastics Machinery & Equipment 7(8), 27, August (1978).©
Fig. 6-35. Mandrel winding unit to make coiled products such as coiled power and telephone cords.

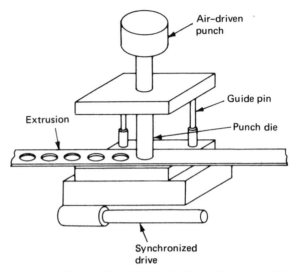

Reprinted from Plastics Machinery & Equipment 7(8), 28, August (1978).©
Fig. 6-36. In-line synchronized punching unit to stamp shapes out of an extrusion on line.

used with thinner and softer extrusions to do the same thing. An example of the use of these punching techniques is the weep hole and nail slot cuts into vinyl siding on line to minimize production costs.

Figure 6-38 illustrates another type of forming operation used to extrude twisted decorative rods from acrylic material. The on-line

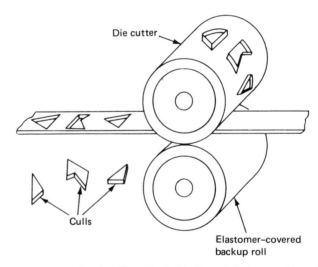

Reprinted from Plastics Machinery & Equipment 7(8), 28, August (1978).©
Fig. 6-37. In-line rotary die cutting unit to cut shaped holes in the extrusions on line.

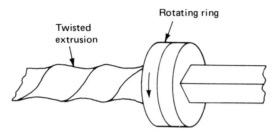

Reprinted from Plastics Machinery & Equipment 7(8), 28, August (1978).©
Fig. 6-38. Rotating ring used to twist an extruded profile.

operations that can be done depend on the skill and ingenuity of the machine designer. Other examples of on-line operations are given in the reference.

Summary

This chapter has covered some of the large variety of downstream units which are needed to produce commercial extruded products. From the large number of examples it can be seen that the equipment removes the extrudate at the correct speed for dimensional control, cools, maintains shape, cuts, winds, and decorates. Other operations are possible on line to make a variety of end products. Essential knowledge to properly design downstream equipment involves a knowledge of variable-speed drives, heat transfer, the characteristics of the materials as they cool and set (commonly referred to as solid-state rheology), and a knowledge of how to synchronize the speeds of the various drive units. From the examples shown, the designer and manufacturing engineer will have some insight into the selection and design of downstream equipment for special applications.

Reference

1. S. Levy, "Process and Equipment for In Line Post Extrusion Forming," *Plastics Machinery & Equipment* **7**(8), Aug. (1978).

Chapter Seven
Coextrusion and Dual-Extrusion Technology

Extrusion machines are basically pumps that deliver supplies of molten polymer materials. The output streams for two or more machines can be combined to make some interesting products that are difficult or impossible to make in any other way. In addition, when coextrusion is a possible manufacturing method, the economics of processing are frequently very attractive as compared with the alternative methods of manufacture of products involving two or more different plastics.

There are two distinct developments in dual material extrusion, and they will be covered in detail since they represent the major advances in the field. Other new developments which stem from these will be discussed briefly to show how the field is expanding to include new equipment and methods for the manufacture of unique products. The two basic areas are the coextrusion of sheet materials and the production of shapes which include two or more materials. Included in the latter category is pipe made by coextrusion methods.

Coextrusion of Sheet and Film

These two materials are considered together since they are essentially different only in thickness. Because the equipment change from the standard methods is basically the same for both, they can be treated as one technology with respect to dual material production. The essential requirement is the layering of the material (either in the die or in a feed-block unit) prior to its entering the die. Both systems are employed in flat-film and sheet extrusion and in blown-film extrusion. The dies and die construction are shown in Figs. 3-11, 3-13, 3-25, and 3-26.

Figure 3-11 shows the in-die combination arrangement for coextruding flat sheet and film materials. The die is equipped with three manifolds, each of which preforms a sheet. The three layers are combined just before reaching the die lips. Since the flow is closely controlled in each of the manifolds and in the approach section to the die lips, there is much greater variation permitted in the material rheology than with the alternative feed-block system. One of the problems with the system is that the die is complex to build and to maintain. In addition, there is a limit to the range of layer thicknesses that make up the sheet built into the die. Drastic alteration of the relationship between layer thicknesses requires changes in the manifold sections to ensure uniformity of sheet layer thickness.

The alternative method for coextruding flat sheet and film is shown in Fig. 3-13. This is the feed-block system. The two resins are combined into a mixed stream in the feed-block unit, and the viscous (nonturbulent, streamlined) flow will keep the layers separate as the materials flow from the feed block through the manifold and out to the die lips. This is shown schematically in Fig. 3-12. The stacked layers in the feed block are reformed to the thin sheet as the material flows through the die structure. There are some significant restrictions on the materials that can be coextruded using the feed-block concept. One requirement is that the materials exhibit similar rheology at the temperature and pressure at the point of combining. Large differences in effective viscosity will result in severe interlayer distortion of the material. Some of this effect is shown in Fig. 7-1 from Ref. 2. This effect is related to the extrusion conditions as well as to the material characteristics. The extrusion rates can be adjusted to minimize the effects.

Figure 3-26 shows a blown-film die using the individual layer formation of the two materials which are combined just before the materials enter the die lips. It is evident that the required die is substantially more complicated to construct than a simple blown-film tool. Analogous to the flat-sheet coextrusion dies there are limitations in the relative layer thicknesses which are built into the die structure. Departing substantially from the designed proportions of the two materials in the film can cause severe restrictions on output rate with the possibility of distorted layers. The dies can be used with materials having very different melt characteristics, and the combining systems have been used to make a nylon–polyolefin combination which would be very difficult in the feed-block system. The melt characteristics of nylon and a material like low-density polyethylene would make streamline viscous coflow almost impossible.

The feed-block system with a blown-film die is shown in Fig. 3-25.

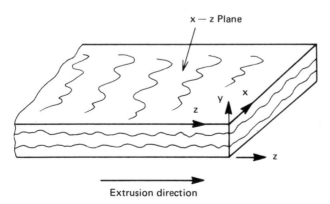

x — z Plane

Extrusion direction

Reprinted from Polymer Engineering & Science 18, 181, Feb. (1978).
Fig. 7-1. Irregularities at interface in coextruded sheet.

The feed-block would be little different from one employed in the flat-sheet and film system. The advantages here are obvious. A standard blown-film tool can be used and the feed block added to the die head. In addition, the relative thickness of the layers is readily changed provided that the melt rheologies of the resins are similar at the combining point temperature and pressure. Another point which is worth nothing is that if more than two streams are combined, the only unit that has to be redesigned is the feed block. The changes in this unit are fairly simple as compared with the die changes that would be necessary for the other combining scheme.

In addition to melt characteristics there are a number of points with respect to materials that need to be considered in choosing combinations that can be coextruded. One requirement is that the materials will adhere to each other. This necessitates similarity in the chemistry of the material as well as materials which can have natural adhesion to each other. One combination widely used is low- and high-density polyethylene. The adhesion is good and the two materials contribute to each other's characteristics. It is equally important that materials be combined that have somewhat similar setting shrinkages. If they are widely different as, for example, a combination of polyethylene and rigid PVC, the film or sheet will have a decided tendency to curl beyond control after cooling. This effect is more serious with thicker layers of materials having high-flex moduli since they would exert substantial bending forces caused by the interlayer shrinkage stresses.

There are several ways in which the adhesive requirements for the material can be met if the materials needed in combination do not have a natural adhesion to each other. One approach is to compound a compatible resin into one layer which will have the required adhesive properties. The use of EVA polymers in polyethylene resins to get adhesion to vinyls is a good illustration of this technique. Another approach is the use of an actual adhesive interlayer extruded along with the other layers of material to make the composite structure. EVA materials, polyvinyl chloride-acetate materials, and some rubber materials have been used in this manner. The specific adhesive layer will depend on the plastics materials used.

To get sheet with uniform layers and a lack of roughness at the interface requires that the selected materials have suitable rheology for the process. In addition, the selection of the specific material viscosity in a given layer is important to the success of the operation. This aspect of the coextrusion process is covered in Refs. 1, 2, 3, and 4. Figures 7-2 a, b, c, and d indicate the differences in flow profiles in several multilayer film extrusions. In each case where there is a sharp discontinuity in the velocities, it is possible to get severe fluctuations at the interface between the layers and a very poor product. It is apparent from this study that the most desirable conditions in terms of film quality are found when the core resin is of higher viscosity than the covering resin. This is especially apparent in the polyethylene–polystyrene system. These illustrations also show that the shear level will have a significant effect on the product.

Figure 7-3 from Ref. 4 shows graphically the results of a study on the changes in the velocity profiles and consequent stability of flow for the system ABS–polystyrene–ABS with changes in skin layer thickness, skin viscosity, total rate, and die gap. From this it is apparent that much can be done to design a successful coextruded sheet system by selecting materials with the appropriate flow properties and by adjusting the die and extruder conditions.

The coextrusion of sheet and film represent a major area of plastics utilization which enables the manufacture of materials with properties that combine the best of several polymers. For example, barrier films can be made with resistance to permeability from both hydrocarbons and water using a combination of polyolefin and polyvinyl chloride materials. Food-contact-grade sheet can be made by the coextrusion of a high-impact polystyrene core with crystal polystyrene skins. Also, since the colors of each skin can be different, there are additional color possibilities for containers thermoformed from the material. Other specific properties can be combined in products, such as scratch resistance for clear materials, surface wetting characteristics for medical applications, and other special uses.

The technology of combinations has been developed with the sheet and film materials, but it can be applied to other composite extrusions.

Coextrusion of Profiles, Tubing, and Other Products

This omnibus classification covers a range of different products and extrusion techniques. Some of the technology is fairly old and is derived from approaches quite different from the sheet-and-film extrusion. The use of a second plastics stream as a marker in wire extrusion, or in the marking of plastics hose and tubing, is a technique that has been practiced for many years and was probably adapted from rubber extrusion methods. The color-coding operation is one that will give the minimum of operational problems in coextrusion since the compounds are generally identical except for the added colorant. The simplest type of die constructions are used in this application.

Another early development in coextrusion is the manufacture of vinyl

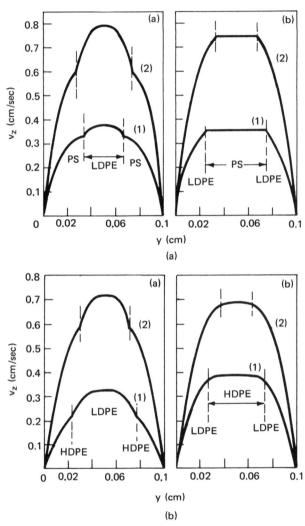

Reprinted from Polymer Engineering & Science 16, 703, Oct. (1976).
Fig. 7-2. Velocity profiles in coextruded sheet material of polystyrene and low-density polyethylene with different layer arrangements and operating conditions. (a) Velocity profiles in the three-layer coextrusion of the PS and LDPE system. (a) PS/LDPE/PS layers: (1) $-\partial p/\partial z = 162.0$ psi/in. (1.12 MPa), Q = 19.4 cc/min; (2) $-\partial p/\partial z = 198.2$ psi/in. (1.37 MPa), Q = 35.2 cc/min. (b) LDPE/PS/LDPE layers: (1) $-\partial p/\partial z = 91.3$ psi/in. (0.63 MPa), Q = 29.4 cc/min; (2) $-\partial p/\partial z = 122.6$ psi/in. (0.84 MPa), Q = 72.4 cc/min. (b) Velocity profiles in the three-layer coextrusion of the HDPE and LDPE system. (a) HDPE/LDPE/HDPE layers: (1) $-\partial p/\partial z = 120.2$ psi/in. (0.83 MPa), Q = 18.4 cc/min; (2) $-\partial p/\partial z = 187.8$ psi/in.

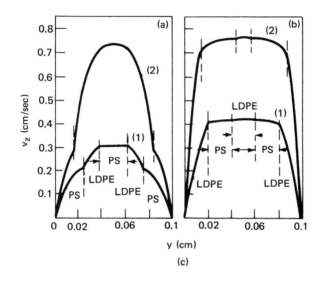

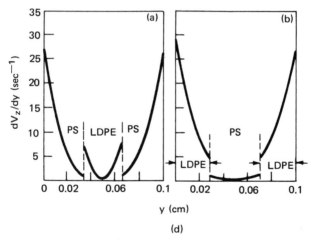

(1.29 MPa), Q = 41.6 cc/min. (b) LDPE/HDPE/LDPE layers: (1) $-\partial p/\partial z$ = 92.3 psi/in. (0.64 MPa), Q = 28.2 cc/min; (2) $-\partial p/\partial z$ = 117.4 psi/in. (0.81 MPa), Q = 64.1 cc/min. (c) Velocity profiles in the five-layer coextrusion of the PS and the LDPE system. (a) PS/LDPE/ PS/LDPE/PS layers: (1) $-\partial p/\partial z$ = 142.8 psi/in. (0.98 MPa), Q = 18.4 cc/min; (2) $-\partial p/\partial z$ = 167.4 psi/in. (1.15 MPa), Q = 32.2 cc/min. (b) LDPE/PS/LDPE/PS/LDPE layers: (1) $-\partial p/\partial z$ = 102.1 psi/in. (0.7 MPa), Q = 21.3 cc/min; (2) $-\partial p/\partial z$ = 143.2 psi/in. (0.99 MPa), Q = 60.3 cc/min. (d) Profiles of the velocity gradient in the three-layer coextrusion of the PS and LDPE system. (a) $-\partial p/\partial z$ = 162.0 psi/in. (1.12 MPa), Q = 19.4 cc/min; (b) $-\partial p/\partial z$ = 95.6 psi/in. (0.66 MPa), Q = 35.3 cc/min.

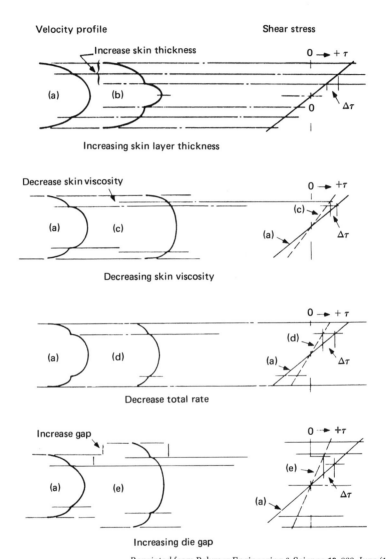

Reprinted from Polymer Engineering & Science **18**, 623, *June (1978).*
Fig. 7-3. Schematic representation of changes in velocity profiles caused by (1) increasing skin layer viscosity, (2) decreasing skin layer viscosity, (3) decreasing total rate, and (4) increasing the die gap.

products employing two different materials having different hardnesses. These so-called dual durometer extrusions are used for a variety of products ranging from plastics sections with attached gasket seals to hinged snap sections which use the softer material as a hinge. The techniques have been and are considered highly proprietary because of the die-design requirements and the material interactions involved in the extrusion. A successful coextruded product is the end result of considerable experimentation which, until recently, was done by a strictly cut-and-try procedure. Recently published papers, including the previously cited references and a paper by N. Minagawa and J. L. White, Ref. 5, have provided an insight into the extrusion conditions in dual durometer extrusion that has made the design of systems more rational and predictable.

In Ref. 5 a system consisting of polyethylene and a filled polyethylene containing titanium dioxide was extruded into a round and a slit die. By varying the relative viscosity of the two materials, and by adding different amounts of the filler, the effect of relative viscosity on the extrusion was determined. The effects of shear rates and other rheological considerations were also examined, and the experiments give a good basis for the prediction of the behavior of coextruded masses. Figure 7-4 shows the three shapes that were coextruded. Figure 7-5 shows the change in material distribution in the extrudate of a round shape with time of extrusion. Figure 7-6 indicates the effect of differences in melt viscosity on the distribution of the material in the round extrusion. With a large difference in melt viscosity there is a major reformation of the coextruded distribution. Figure 7-7 shows the effect of the different melt viscosity ratios on a thin-face interface flat-strip coextrusion, while Fig. 7-8 shows the same range of viscosities effect on a side-by-side coextrusion. In the reference, the effects of die land length and other variables on the material distribution are discussed.

It is apparent from these studies that the best condition for making coextrusions is to have the melt viscosity of both resins as close together as possible at the point of stream combining. This will result in the minimum disturbance at the interface between the two materials and

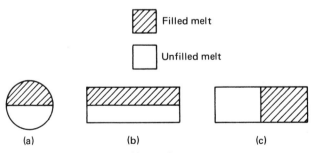

Reprinted from Polymer Engineering & Science 15, 827, June (1978).
Fig. 7-4. Cross section and initial interface of (a) capillary, (b) slit with interface parallel to long dimension, and (c) slit with interface parallel to short dimension for coextrusion of filled and unfilled polyethylene.

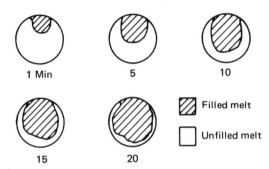

Reprinted from Polymer Engineering & Science **15**, 827, Dec. (1975).
Fig. 7-5. Interface shapes and phase distribution in capillary as a function of extrusion time.

will produce the best product. This condition can be met several ways. One is to use materials which have been formulated to produce substantially identical viscosities at the melt temperature and shear rate that will be found in the coextrusion die. Sometimes this is difficult to do. An alternative method of operation is to select materials which are reasonably close in rheology and to adjust the viscosity of one material or the other by changes in either temperature or shear history. This can be done by adjustments in the extruder operating conditions. If the two melt streams must be at different temperatures for proper combining conditions, it complicates the die by requiring that the materials be maintained at different temperatures while being combined in the tool. A good part of the art in coextrusion of shapes is in the design of the dies to combine

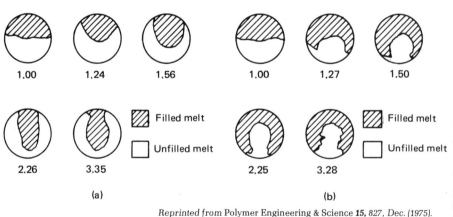

Reprinted from Polymer Engineering & Science **15**, 827, Dec. (1975).
Fig. 7-6. Interface shapes for capillary dies as a function of viscosity ratio: (a) viscosity filled > viscosity unfilled, (b) viscosity filled < viscosity unfilled.

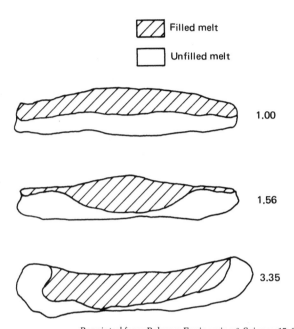

Reprinted from Polymer Engineering & Science **15**, *828*, Dec. (1975).
Fig. 7-7. Interface shapes for rectangular slit die with initial configuration parallel to slit length as a function of the viscosity ratio.

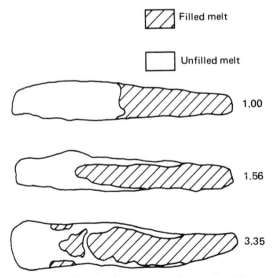

Reprinted from Polymer Engineering & Science **15**, *828*, Dec. (1975).
Fig. 7-8. Interface shapes for rectangular slit die with initial configuration parallel to the short dimension as a function of viscosity ratio.

the streams in the proper manner, and to control the rheology of the melt to avoid rough interfaces and poor adhesion.

The other consideration in the coextrusion of shapes is the adhesion factor. This can be more of a problem than it is with sheet coextrusion. Frequently, there is only a small interface between the coextruded materials, and this places a premium on good adhesion. This is sometimes supplemented by making a corrugated or interlocking interface to improve the holding capability between the different sections of the coextrusion. Adhesive interlayers are also used to improve the holding power at the joints.

A typical die for coextrusion which illustrates one of the classes of coextrusion dies is shown in Fig. 7-9. This die is a covering die where one of the materials is extruded around the other. A typical use of this type is to cover a material with fair to poor weathering properties with a covering resin having excellent weathering characteristics. The die has a straight-through section for the mainstream of material. The coextruded stream is introduced through a side port partway through the land length.

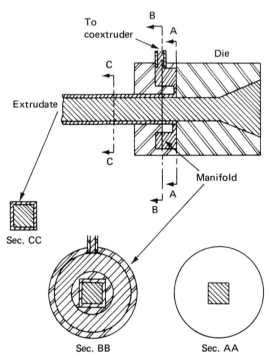

Fig. 7-9. Schematic of a simple profile coextrusion die showing manifolds and stream-entry points.

Typically, the stream is fed into a manifold section which distributes the material around the die orifice, and the material then flows from the manifold across a controlling secondary land into the main die land section. The design of the manifold and land is one of the critical elements in a successful coextrusion die.

The die shown is a useful starting point to discuss the mechanics of die design for coextrusion of shapes. In each case the die will have a central stream or main extrusion stream and entry ports for other streams. Depending on the shape and the extent of coverage of the primary stream by the coextruded streams, the manifold(s) may or may not wrap around the primary stream. In any event, the manifold will have a considerably enlarged flow volume as compared to the entering secondary stream in order to permit pressure balancing to the lands which control the flow to the main die land section. The flow to the die land section will be the result of the pressure from the extruder and the pressure drop across the secondary land connecting the manifold with the main die land. This pressure drop is a direct function of the land length and is inversely proportional to the cube of the land depth. By changing the opening or the length of the land, the flow to the main die land can be locally controlled to give the desired relative flow of material to make the coextruded shape. In the example of Fig. 7-9, a uniform cover on the main extrudate is desired, and the land lengths will have to be adjusted to produce uniform flow along the entire secondary land.

Three basic dual durometer-type coextrusions are illustrated in Fig. 7-10. The first is a covering operation. The tool for this was discussed in the previous section. The second one is a twin section which has comparable amounts of two different resin streams. An example of this type of extrusion is a squeegee unit which has a large flexible blade coextruded onto a rigid support handle. This type of extrusion has been used in a standard auto windshield tool. The third one is the small add-on extrusions. Small amounts of material can be added for several purposes. One use is for introducing a gasket element into a rigid part. Another application could be the use of a magnetic material to make the extrusion hold magnetically to an iron substrate. Since the secondary streams are usually small and discrete, the die design will be different from the ones used for the other types of coextruded sections.

An example of a two-stream extrusion is given in Fig. 7-11. The part is a special seal section with the square corner sections of rigid material and the plus (+) shape section of a soft-seal compound which will make a four-way seal. One of the important dimensional requirements for this part is the squareness of the total cross section. This makes the tool design difficult since the shaping requirements of the profile and the

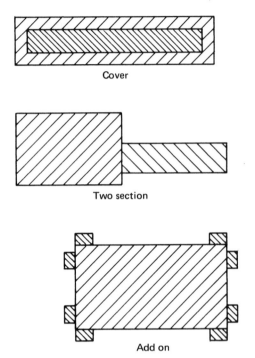

Cover

Two section

Add on

Fig. 7-10. Illustration of typical coextrusion cross sections.

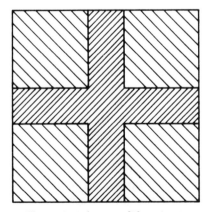

Fig. 7-11. Illustration of some solid section coextrusions.

relative flow control are not achieved by the same corrections. A sketch of a die design is given in Fig. 7-12. As with the covering die there is a manifold all around the primary stream. The rigid center shape is made in the first section of the die before the second stream is introduced. Since the secondary stream is the rigid material, it is important that the pressure balance be made carefully. Otherwise the softer material can be squeezed out between the rigid, square corner sections. In order to produce the desired result it would be necessary to have a material of low-melt viscosity to make the rigid corner sections. The flow into the main die land would be controlled by adjusting the length and depth of the secondary land section so that the flow would correspond to the required fill rate. A long land past the point of introduction of the secondary melt stream is necessary to help form the section.

An important class of add-on extrusions is the gasketed rigid section. An example of this is shown in Fig. 7-13. Sections similar to this are used in PVC window profiles and in the gasket sections used in appliances. The two small sections of the profile shown in the darker shading are soft materials which will gasket a glazing element inserted between them. The die used to coextrude this section would be designed in the manner shown in the sketch in Fig. 7-14.

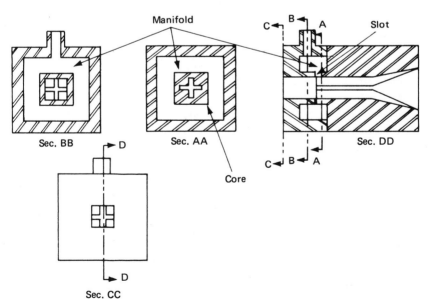

Fig. 7-12. Schematic drawing of a coextrusion die to make an embedded cruciform shape (see Fig. 7-11).

Fig. 7-13. Coextruded gasketed section profile.

This die is constructed differently from the two previous ones. It does not have a typical manifold section. Instead, the secondary melt stream is introduced into the main die land by means of two runnerlike secondary lands. In the other cases, the melt-flow control will be determined by the length and depth of these secondary lands, but now each add-on area can be separately adjusted to produce the required profile. The other difference between this die and the previous examples is that the main extrudate section is a hollow unit. This requires a little longer land section and good control of the secondary melt streams to prevent the collapse of the hollow sections.

The coextrusion of foam with solid material for a cover is another important application for the processes. This is done with sheet and film as well as with shapes. The die construction used is shown in Fig. 7-15. The die is very similar to that used for the hollow section coextrusion except for the introduction of the foam compound into the interior of the skin section. The vacuum sizer used is similar to the one used to make the hollow extrusion. For a product such as a window profile the foam would improve the structural stiffness of the product and increase the buckling and bending strength.

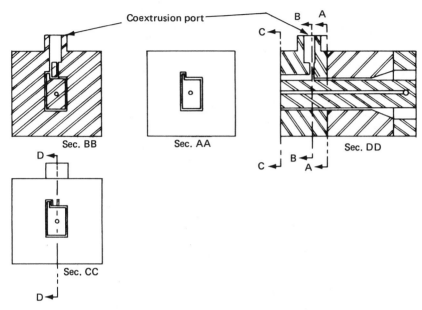

Fig. 7-14. Profile extrusion coextrusion die to make profile of Fig. 7-13.

Figure 7-16 illustrates the use of three melt streams to make a product. The window section of Fig. 7-15 has a gasketing section of flexible compound added to the extrusion. The manifold arrangement used is similar to that described for the other gasket coextrusion. The operation is more complicated with the three melt streams since they must all be balanced to give the required material proportions to fill out the section. Proper material selection and die design are required to make an

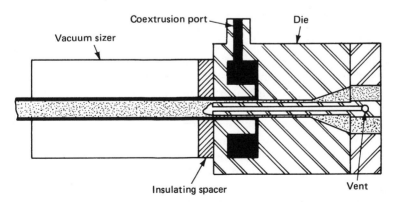

Reprinted from Plastics Machinery & Equipment 6(11), 12, Nov. (1977).©
Fig. 7-15. Coextrusion die to make foam-cored solid skin coextrusion.

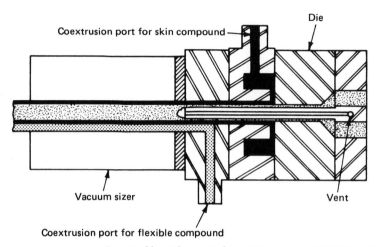

Reprinted from Plastics Machinery & Equipment **6***(11)*, 12, *Nov. (1977).*©
Fig. 7-16. Gasketed profile with foam core and coextrusion die to make the coextrusion.

operational system. During test trials it is generally necessary to make adjustments in the secondary lands to balance the flows.

Plastics pipe and tubing made by coextrusion result in some unusual products. For example, flexible polyvinyl chloride lined with urethane has applications in the handling of abrasive slurries and in blood handling for medical devices. Another combination is flexible vinyl lined with polyolefins for solvent carrying lines. In pipe, the combinations would include polypropylene lined with PVC for carrying oxidative chemicals and one of the most significant in terms of volume of product made, ABS pipe with a foam layer between the inner and outer walls of the pipe. The cost and weight savings that result from the foam layer has enabled this material to be a major factor in DWV (drain waste vent) piping for waste lines.

A straight-through design for coextruded pipe or tubing is sketched in Fig. 7-17. In the illustration the coextrusion inlets are shown as independent ports so that the die can be arranged to produce a pipe with different materials for the inner and outer walls. The main stream or center streams is the straight run through the die. This die can be used to make a foam center coextruded pipe as well as a double or triple extrusion solid coextruded pipe section.

The outer layer is applied in a manner similar to that used for the other coextruded shapes discussed. An annular manifold feeds the covering material around the main extrusion stream. If the center coating detail were left out of the die, this would be the die construction for a two-layer pipe or tube such as the flexible PVC–polyolefin combination

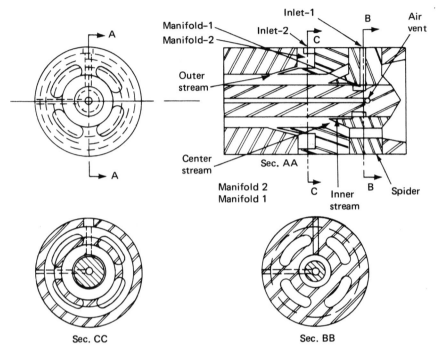

Fig. 7-17. In-line pipe extrusion die to make coextruded pipe with solid skins and foam inner wall.

that was mentioned as a coextruded product. In order to coat the inner wall of the tube it is necessary to carry the polymer melt through one leg of the spider supporting the central core pin in the die as is shown in Fig. 7-17. The manifold to distribute the material around the tube is cut into the center pin in the illustration. This is the most convenient method of construction, although it may weaken the pin. An alternative arrangement would be to undercut the supporting spider section. A die land is formed between the pin and the spider support to control the flow of material to form the center lining of the tube.

As an alternative construction, it is possible to omit the outer coating detail and to make a tube which is internally coated through manifold 1 and inlet 1. This technique is used when very thin inner layers are placed into tubes in a straight-through die such as the one shown. Better control of the inner coextruded layer thickness is possible with this arrangement. A product made using this combination of streams is the urethane-lined flexible PVC material used for medical and abrasion resistant applications.

In a product such as the DWV drainage pipe, with the foam center in the wall manifolds, 1 and 2 both have the same ABS material flowing in them since the two entry ports (1 and 2) would be connected to the same extruder. In most cases the inner and outer wall layers will be of about the same thickness, and their combined thickness is usually in the range of 10–25% of the total wall thickness. The core material would be the same resin as the skins, with the addition of a blowing agent to reduce the density to about one-third that of the skin materials. The weight reduction per foot with substantially the same performance is about one-half that of a solid wall pipe.

The relative flow of the secondary streams to control the wall thickness is done by adjustment of the land lengths which connect the secondary manifolds with the center stream.

The spider has entry holes to the center pin along two legs. One hole connects the center vent to the outside air. The other entry hole is used to carry the resin from inlet 1 to the manifold 1. The die construction is somewhat intricate, but it is easy to assemble and to disassemble for cleaning.

Crosshead arrangements can also be used to make dual and triple coextrusions. The constructions would be similar, and the main advantage might be the elimination of the spider feed for the center stream. This would be done at the expense of easy radial flow balancing. The general trend has been to the use of straight-through dies to minimize the complications of roundness, especially when foam is used as one of the layers.

The sizing of foam-core pipe is somewhat more complicated than the direct sizing of standard pipe. If the sizing is done by a vacuum sizer unit, the bore will generally be much less accurate and consistent than for solid pipe. If an internal cooling mandrel is used to size the pipe, then the outer diameter will vary substantially. Where accurate bore and OD are required, both the vacuum sizer and the mandrel cooling are used. The foam will adjust the wall of the pipe to meet both dimensional requirements.

A solid embedment coextrusion in Fig. 7-18 is shown with several internal streams. The major application for this type of product is decorative, such as striped acrylic sections. In many instances the internal streams can be functional. For example, the internal stream could be a conductive plastics compound which would be used for high-voltage low-current applications. Magnetic materials for use in an electronics device is another possible internal stream material.

Figure 7-19 shows the die construction for making the product shown in 7-18. The approach is somewhat similar to the internal stream supply used in the coextruded pipe with the internal stream feed. In this case,

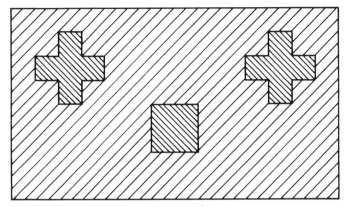

Fig. 7-18. Cross section of embedment coextrusion.

however, all of the streams are introduced through a manifold section which is attached to the die structure and is opened to allow the mainstream to flow around the delivery nozzles. Design of the nozzles would follow the methods used for individual extrusion dies. The exterior part of the nozzles will also have to be shaped to blend the outer mainstream to the shape issuing from the internal nozzles. The application for this method would be for coextrusions with small multiple internal streams in a large cross section of the embedding resin.

Summary

The combination of several materials by coextrusion is useful in the manufacture of a wide range of products ranging from sheet and film to pipe and profiles. By combining two or more materials it is possible to

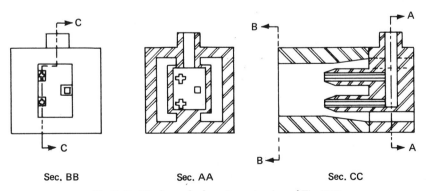

Fig. 7-19. Die for embedment coextrusion of Fig. 7-18.

obtain cost-effective products that cannot be made from a single material. The art involved is a combination of die design, machine control, and material selection. The key to successful coextrusion is compatible melt rheology of the materials combined and in most instances, good intrinsic adhesion between the materials. The latter can be brought about by the use of interlayer adhesives which are introduced as additional coextruded streams. The die designs require ingenuity in bringing the melt streams into the proper relationship and in combining them in the tool. Tools for the production of coextrusions shown are useful as a guide to the design of tooling for many new coextruded products.

References

1. C. D. Han and R. Shetty, "Studies on Multilayer Film Coextrusion I—The Rheology of Flat Film Coextrusion," *Polymer Engineering & Science* **16**(10), 697, Oct. (1976).
2. C. D. Han and R. Shetty, "Studies on Multilayer Film Coextrusion II—Interfacial Instability in Flat Film Coextrusion," *Polymer Engineering & Science* **18**(3), 180, Feb. (1978).
3. C. D. Han and R. Shetty, "Studies on Multilayer Film Coextrusion III—The Rheology of Blown Film Coextrusion," *Polymer Engineering & Science* **18**(3), 187, Feb. (1978).
4. W. J. Schrenk, N. L. Bradley, T. Alfrey, Jr., and H. Mack, "Interfacial Flow Instability in Multilayer Coextrusion," *Polymer Engineering & Science* **18**(6), 620, June (1978).
5. N. Minagawa and J. L. White, "Coextrusion of Unfilled and TiO$_2$ Filled Polyethylene: Influence of Viscosity and Die Cross Section on Interface Shape," *Polymer Engineering & Science* **15**(12), 825, Dec. (1975).

Chapter Eight
Extrusion of Cellular Plastics Products

The extrusion of cellular or foamed plastics is an interesting and important aspect of the plastics extrusion art. The introduction of gases into a plastics material makes possible new materials with unique properties. Among the other improvements is material cost reduction made possible by replacing some of the resin with a low-cost gaseous filler.

There are several subjects of interest in this field. One is the materials and blowing-agent technology used. Another important topic is the equipment required to operate with specific materials and their corresponding blowing agents. Finally there are the special tooling requirements for extruding foam products which range from sheet and tube to profiles.

Blowing Agents and Foaming Processes

Two fundamental types of blowing or expanding agents are used to make cellular plastics products by extrusion. Chemical blowing agents, which are complex chemical materials that decompose in a well-defined temperature range to release large quantities of gases (usually nitrogen, carbon dioxide, or hydrogen), can be used. A list of these blowing agents is given in Table 8-1. By the selection of a suitable agent with the appropriate decomposition temperature range and appropriate modifiers, foaming can be done with a large number of polymer materials.

The second type of blowing agent is the solvent blowing agent. This is a material which is a gas at room temperature and is soluble in the molten polymer. The use of this type of blowing agent requires the use of specially modified extrusion equipment to enable it to be introduced into the extrusion machine under pressure. Table 8-2 lists several of the currently used solvent blowing agents.

Each of the blowing agents operates with a variety of resins. There are other constituents required in each of the material systems. In both cases nucleating agents are necessary to produce fine uniform cell sizes. These materials, typified by talc and finely divided calcium carbonate, act as starting locations where the dissolved blowing agent gases can come out of solution. Because they are not effectively wetted by the resins, the surface tension or interfacial tension is lowered. This permits easier release of the gas from solution and reduces the pressure on the mass of material.

Other materials can be added to the chemical blowing agents' (CBA)

Table 8-1. Chemical Blowing Agents for Extruded Foam

Name	Temperature °F (°C)	Gas yield (cc/g)	Gases
Azodicarbonamide (ABFA)	428 (220)	220	N_2, CO_2, CO, NH_3
4'-Oxybisbenzene sulfonyl hydrazide (OBSH)	302–374 (150–190)	125	N_2
p-Toluene sulfonyl semicarbizide (TSSC)	379 (193)	140	N_2, CO, CO_2, NH_3
5-Phenyltetrazine	450 (232)	200	N_2
Trihydrazinatriazine (THT)	509–554 (265–290)	175	N_2, NH_3

recipe to alter the temperature of the onset of decomposition. This permits some adjustment to match the blowing temperature to the melt rheology of the polymer used. The literature from the suppliers contains a great deal of specific information on the recipes for the blowing agents used with the resins, ranging from polyethylene to polycarbonate. Specific recipes require testing in the operation to ensure proper blowing.

The process requirements for using CBA are basically simpler than those for the solvent blowing system. The resin and other components for the blowing agent system can be combined either by dry mixing or by the hot compounding methods. In hot compounding it is important to avoid premature blowing of the resin mass. The hot compounding should be done at a temperature well below the blowing temperature for the CBA. In the case of vinyl foams, the present approach is to use a dry blend which minimizes the risk of preblow. In the case of polyolefins and polystyrene, either a dry mix or a concentrate containing the blowing agent is the most convenient method of incorporating the CBA. Since the amount added is relatively small, usually less than 10% of the mix, it is adequate to use a mixing aid such as a white mineral oil with the dry mixing technique.

The resin mix is fed to the extruder and extruded at a suitable melt temperature to produce complete decomposition of the CBA in the machine barrel. When the product is exited from the die, it expands to form a cellular material. There are several important limitations in the use of CBA. For one thing, the residue of organic compounds left after

Table 8-2. Solvent-Type Blowing Agents

Isopentane
Fluorocarbon-11 (monofluorotrichloromethane)
Fluorocarbon-12 (dichlorodifluoromethane)
Carbon dioxide
Nitrogen
Ammonia

decomposition can have an adverse effect on the physical properties of the product. Another problem is that the amount of blowing agent that can be added is limited. This, in turn, limits the minimum density that can be achieved. Usually the density of foams made from CBA is limited to those above 0.25 g/cc.

The foams that are made with CBA are good, fine-cell materials if the proper processing temperatures are held. In addition, the techniques used are broadly applicable to many resins and involve a minimum of equipment complexity. Products such as furniture and wall moldings, as well as picture frames, are made by the extrusion of CBA foams at good production rates. Another advantage of the CBA-based systems is that good formulations will be self-skinning and produce good dense skins on the extrusions.

The solvent-blown-foam systems are much more complicated to operate. In these applications it is necessary to inject the liquid blowing agent into the resin mass after it has been completely plasticated. In order to do this, it is necessary to inject the blowing agent under high pressure [usually from 1000 to 5000 psi (6.89 to 34.45 MPa)] using a metering pump to control the amounts relative to the resin flow required to produce foams of the desired density.

Nucleating agents are required for the solvent blown systems as they are for the CBA systems. They are probably more critical in the solvent systems since the residues from the decomposition of the CBA materials can act as nucleating centers while this is not true for the solvent resin mixes. The proper concentration of the correct nucleating agent is essential for fine cell solvent blown foams. If the foams are to be low-density materials, the concentration of nucleating agents is directly related to the concentration of the blowing agent used.

One of the major reasons for using solvent blown foams is that it is possible to produce very-low-density materials by this technique. Generally, solvent blown foams are made to a density of 0.2 g/cc or less and commercial polystyrene foams are made by solvent blowing with densities as low as 0.015 g/cc. Polyolefin foams have slightly higher low-density limits. This is related to the solubility characteristics of the resins as well as to the orientability characteristics.

An interesting point should be noted with respect to density and physical properties. When the dissolved blowing agent is released from solution, the effect is to reduce the melt temperature. As a result, under optimum foaming conditions the gas is orienting the resin as the cells form. This has two effects. One is to restrict the expansion of the cell as the polymer structure becomes cooler and more oriented. The other effect is to make the foam structure stronger than expected as a result of the orientation produced by the blowing. The orientation effect must be taken into account in the formulation of the resin mix. The additional pressure needed to expand the oriented polymer requires either more blowing agent or one with a higher vapor pressure at the blowing temperature.

The most common combinations used in making extruded solvent blown foam are isopentane or fluorocarbon 11 with polystyrene, and fluorocarbon 11 with polyethylene. Some fluorocarbon 12 is used to give higher pressures for blowing, and combinations of F-11 and F-12 are widely used. Nitrogen and carbon dioxide are generally not as convenient, and, consequently, they are not widely used.

Equipment Requirements

Foams made by the CBA method require an extruder of conventional design with a good temperature control system and preferably with a barrel L/D of 28:1 or longer. The screw design should be selected based on the material and the form of the material fed to the machine. For the extrusion of polystyrene foam, a polystyrene screw with a high compression ratio and long metering section is used. The screw should be equipped with a mixing section for best results. In the case of rigid polyvinyl chloride foams made from dry-blend powder, the screw should be a medium compression screw to take into account the low-bulk density of the powder blend. The metering section should be long and fairly deep in order to minimize frictional heating effects. A pin-mixing section on the screw may be advantageous in obtaining good mixing for fine foams at high production rates.

In the case of the polystyrene foam, the feed hoppers should be able to handle the material to minimize segregation of the blowing agent. The dry-blend powder may need a special hopper with steep sides to prevent bridging in the hopper. Some dry-blend formulations may need the assistance of a hopper agitator unit to reduce the tendency for bridging. Occasionally a crammer feeder may be necessary to get uniform material feeds.

The head and breaker plate assemblies used with the foams are the same as those used to extrude similar products from solid plastics. It is necessary to avoid regions where resin can hang up and decompose. This is more important for this operation than for solid extrusion since material held too long will overblow and create poor-quality foam. In addition, when extruding PVC foams the hydrochloric acid that results from decomposition of the resin can activate the CBA materials excessively. This will lead to blowouts of material from the die caused by the excessive blowing agent pressure.

The solvent blown foams require a more involved equipment package than the CBA-based foams. The reason for this is that the blowing agent must be pumped into the extruder at a strategic location under high pressure. In addition, the blowing agents act as a softening agent on the polymer causing the melt viscosity to decrease drastically with the addition of the blowing agent. As a result, the melt must be cooled enough to increase the melt viscosity to the point where the cells formed by the gas release will not rupture. This cooling can be done several ways and the equipment packages to make the foam are built accordingly.

One of the equipment packages is shown schematically in Fig. 8-1. In this scheme two tandem extruders are used. The first machine is used to plasticate the material. The plasticated material is fed through a melt connection into a second, larger machine which acts as a melt pump. At the last section of the plasticating machine the blowing agent is pumped into the barrel where it is mixed with the polymer. The screw in the plasticating machine is equipped with a mixing section past the injection point to mix the blowing agent into the resin.

The melt pump unit is a larger diameter machine with a deep flighted screw. As a result, the second machine operates at much lower screw speeds and with a minimum of shear on the melt. This prevents shear heating effects from overcoming the cooling which is the purpose of the second machine. The barrel is operated at a low enough temperature to reduce the melt temperature to the point where the necessary melt viscosity is achieved for good foam formation. The dies are attached to the exit section of the second stage or pumping and cooling machine.

There is one feature of the screw in the plasticating unit common to all single-screw foam packages. It is the blister section shown in Fig. 8-2. This restriction is necessary to form a thin section in the melt to act as a seal to prevent the blowing agent from escaping upstream through partially plasticated melt through the feed hopper. The blister section on the screw will be located slightly upstream from the injection port for the blowing agent.

Another feature common to all of the solvent blowing systems is the use of a check valve at the point of injection of the blowing agent into the machine barrel. The design of a typical check valve is shown in Fig. 8-3. The purpose of the check valve is to prevent polymer from escaping from the machine barrel and entering the injection section that connects to the blowing agent supply in the event of loss of blowing-agent pressure.

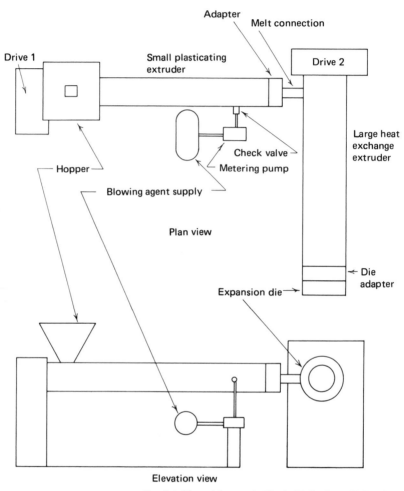

Reprinted from Advances in Plastics Technology *1*(1), Jan. (1981).
Fig. 8-1. Tandem extruder injection foam line for solvent-blown plastics foam.

There are three requirements for the valve that are important. It should act rapidly in the event of pressure loss, it should be resistant to clogging from the polymer, and it should be simple and easy to clean in the event of a backflow.

Other arrangements can be used in a solvent blown-foam system besides the tandem extruders. The essential requirement of the extrusion unit is that it be capable of mixing in the blowing agent and then having the solvated mass cooled to the proper temperature for blowing. In one system illustrated in Fig. 8-4, this is done by using a very long extruder

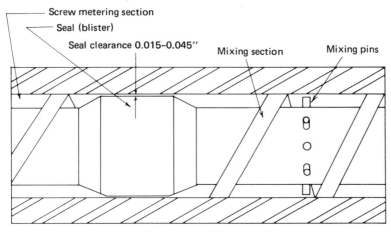

Reprinted from Advances in Plastics Technology *1*(1), Jan. (1981).
Fig. 8-2. Blister seal section on screw for single-screw extruder injection solvent-blown plastics foam.

with a deep flighted section in the cooling section of the machine, but the equipment has poor throughput for the screw size. This results from the fact that the screw speed is the same in the entire length of the process so that there is appreciable shear heating of the melt. The screw can be made with deep flutes and high helix angles and, in some instances, with twin-fluted screws to minimize the shear heating effect. The speed of the screw will limit output of cooled foam.

Another single-machine system makes use of a static mixer in the

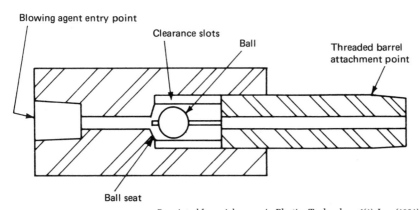

Reprinted from Advances in Plastics Technology *1*(1), Jan. (1981).
Fig. 8-3. Ball check valve design for solvent injection port on solvent-blown plastics foam extrusion line.

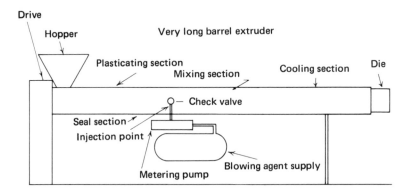

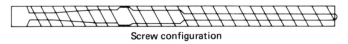

Screw configuration

Reprinted from Advances in Plastics Technology *1(1), Jan. (1981).*
Fig. 8-4. Extended barrel solvent-injection plastics extrusion foam line.

heat-exchange portion of the process. The static mixers will not impart any significant shear-heating effect to the melt, and the intimate mixing action with several passes of the melt against the barrel wall results in effective cooling of the melt in six to eight barrel diameters (Fig. 8-5). This has been effectively done in small [2 and 2½ in. (5.08 and 6.35 cm)] machines. The problem is not solved as effectively with static mixers in larger machines. The reason is that the heat-transfer area increases in proportion to the increase in barrel diameter, while the amount of material to be cooled increases as the square of the barrel diameter. Since the barrel wall is the controlling heat-transfer surface, longer lengths of the barrel are required for larger machines and the number of diameters of static mixer is substantially increased. With the same throughput rate, doubling the diameter will necessitate increasing the static mixer length by two times. The result is that the machine will need 12 to 18 diameters of static mixer for adequate rates of cooling.

Another approach to the problem is the use of special heat-exchanger units attached to the extruder head. Such a unit is shown in Fig. 8-6 taken from U.S. Patent 4,222,729. By using a heat-exchange block with radial polymer passages, it is possible to achieve high rates of heat removal in a short space in the machine. The heat exchanger can be used alone or with a short static mixer to minimize temperature variations. Other types

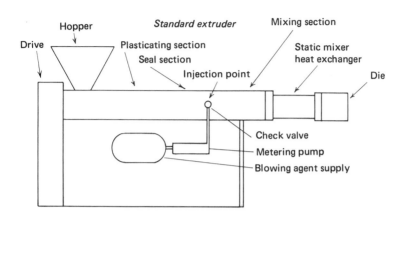

Barrel/screw configuration

Reprinted from Advances in Plastics Technology **1**(1), Jan. (1981).
Fig. 8-5. Solvent-injection plastics foam extrusion line using a static mixer as a heat exchanger.

of heat exchangers such as the tube type have been employed with some of the polyolefin materials where the clean up is not difficult.

There is some usage of twin-screw machines to produce foam plastics because of their low shear characteristics. By using a portion of the machine operating length for plastication, and the remainder after the blowing agent injection port for cooling, the twin-screw units can contain the process in one machine. Figure 8-7 illustrates such a package. The point of entry into the barrel is shown and the machine is a counterrotating screw unit. The desirable length of such a machine is about 20 L/D or more in order to get sufficient distance for plastication and cooling. One detail that is a problem with the twin-screw machines is a seal on the screws which serves the same function as the blister shown in Fig. 8-2. Such seals have been made but the structural details are not available.

In order to clarify the equipment requirements it is useful to have a complete flow chart for the steps in the solvent foam process. This is shown in Fig. 8-8. All of the systems of equipment described have each of the elements shown in the flow chart.

The one unit common to all of the systems is the injection pumping system for the blowing agent. A schematic of a typical unit is shown in

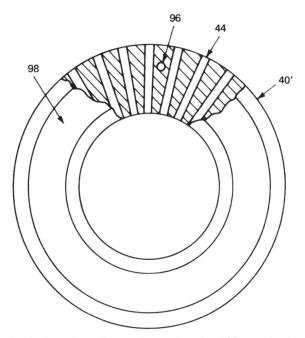

Fig. 8-6a. Single plate of plate heat exchanger for solvent-blown extrusion foam line.

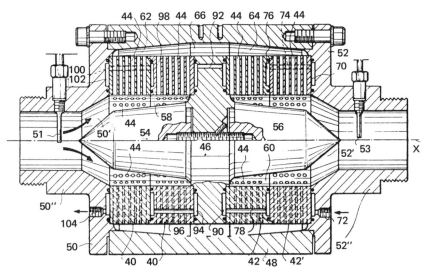

Fig. 8-6b. Assembled plate heat exchanger diagram for use on solvent-blown foam
extrusion line.

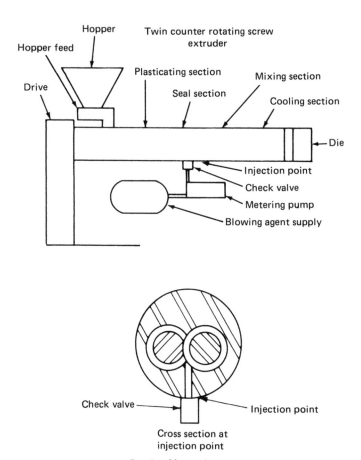

Reprinted from Advances in Plastics Technology 1(1), Jan. (1981).
Fig. 8-7. Solvent injection extrusion foam system using a twin screw extruder.

Fig. 8-9. The unit includes storage tanks for the blowing agent with a pressure intensifier to deliver the agent to the metering pump. The metering pumps generally used are oil-filled diaphragm pumps of the piston activated type which are capable of generating pressures up to 10,000 psi (68.9 MPa). The rate of delivery of the pumps can be controlled by adjusting the pump stroke or by varying the pump speed. The practice is to use the stroke adjustment for the range setting and to adjust the motor speed to set the actual delivery rate of the blowing agent. To complete the assembly requires pressure gages to read head pressures and inlet pressures, check valves to minimize backflow, a flow meter to check on delivery rates, and a safety rupture disk assembly for safety in

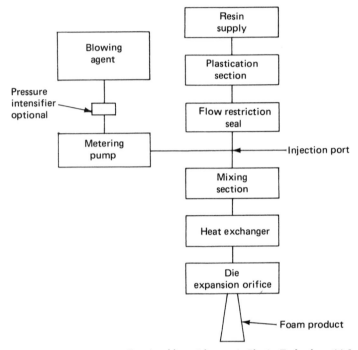

Reprinted from Advances in Plastics Technology 1(1), Jan. (1981).
Fig. 8-8. Flow chart for gas-injection solvent-blown extrusion foam process.

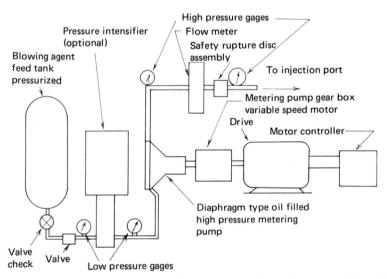

Reprinted from Advances in Plastics Technology 1(1), Jan. (1981).
Fig. 8-9. Solvent injection feed system layout for solvent-blown foam.

the event of excessive pressure buildup in the blowing-agent lines due to blockage.

With slight variations, the system is usable for isopentane, fluorocarbons, and carbon dioxide. In the case of the isopentane with its explosion and fire hazards, the entire system, including the extruders, has to be made explosion proof. Using liquid carbon dioxide requires some changes in the system to insulate it thermally to prevent excessive evaporation pressures in the feed system to the pumps.

Dies and Downstream Equipment

One of the major foam products is thin sheet made from polystyrene and polyethylene. The usual method of manufacture of low-density sheet is to extrude it through a tubular die. The expansion of the foam causes the tube to grow in diameter as well as in length and thickness. As a result, the material comes off the die like a blown film. This is shown in Fig. 8-10. In a low-density foam in the range of 1–2 lb/ft^3 (16–32 kg/m^3) the expansion in diameter is four to six times, and this is clear from the picture. Figure 8-11 shows a complete line for making expanded styrene foam sheet. Included are the two tandem extruders equipped with gas

Fig. 8-10. Extrusion of solvent-blown polystyrene foam sheet.

Fig. 8-11. General view of extrusion line for extruding solvent-blown extrusion foam sheet.

injection apparatus, a large-diameter tubular die equipped with a splitter unit, the sheet-flattening unit, the sheet-handling units, and the winders in the foreground to collect the product. The front of the die as shown in Fig. 8-12 shows the exit opening for the die. As a rough rule of thumb, the opening is about one-half the thickness of the sheet allowing for expansion and drawdown effects. Figure 8-13 is a closeup view of the die from the rear, and a clear picture of the splitter which is used to make two sheets from the tube.

In most of these lines the cooling is accomplished with air from blowers. The major cooling and setting is internal in the foam. The expansion and release from solution of the blowing agent will cool the material to a set shape. The additional cooling is used to remove the residual heat.

The internal die configurations used in the tubing dies for this application are similar to the ones used in blown film dies. The essential requirement is uniform distribution of the material around the die to ensure uniformity in the extruded sheet. The spiral distribution method has been extensively used in both polystyrene and polyethylene foam sheet extrusion.

Heavy sheet and block foam are usually made on a line with a sheet die to control the extruder output. The resulting foam is cooled and set

Courtesy Welex Co.
Fig. 8-12. View of tandem extruders and die for solvent-blown sheet extrusion.

Courtesy Welex Co.
Fig. 8-13. Close-up view of circular die used for extruding solvent-blown extruded sheet.

between two conveyor belts as shown in Fig. 8-14. The large expansion effects as the materials come from the die can cause undesirable variations in density as the block expands. Modifications are made in the choker bar and die lips in the sheet die to minimize this effect, but details of commercial systems are not available. One method that has been proposed to take care of this effect is to use a set of convoluted die lips to allow better free expansion of the foam.

Figure 8-15 shows an additional unit attached to the system to apply vacuum to the extrudate in order to reduce the density by reducing the pressure against which the blowing agents expand the resins. This can help reduce the density by as much as 50% and is helpful in low-density packaging material production. Figures 8-14 and 8-15 show the two general methods of handling the extrudate. The cooled conveyor in 8-15 and the cooled sizer in 8-14 are intended to set the outer layer of the foam block. Final cooling is generally done by storage of cut sections between cooled platens.

One of the other products made from foam is tubing or pipe which can be used as thermal insulation. The presently used materials for this application are flame-retardant polystyrene and polyethylene. The tooling design is a simple thin-wall tubing die from which the material is extruded into a sizer unit. Since the foam will expand in all directions, the usual practice is to make the diameter somewhat smaller than the desired outside diameter of the tubing needed. The outside diameter is controlled by means of a vacuum sizer sleeve similar to the ones used to size pipe and tube in solid materials. In some cases the inside diameter is left to a free expansion size, which can be controlled to some extent by the drawing rate from the puller. This leaves a rough interior in the tubing which is acceptable for some applications, but not all. The alternative is to add an internally cooled sizing mandrel such as the ones used for internally sizing tubing. This will set the inside diameter to a

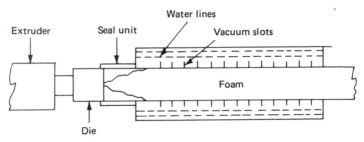

Reprinted from Plastics Machinery & Equipment 8(1), 15, Jan. (1979).©
Fig. 8-14. Sketch of extrusion arrangement for extruding plastics solvent-blown foam insulating stock using sizer.

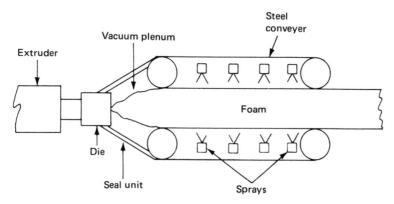

Reprinted from Plastics Machinery & Equipment 8(1), 15, Jan. (1979).©
Fig. 8-15. Sketch of extrusion arrangement for extruding plastics solvent-blown foam using steel conveyor sizing.

predetermined size with good surface finish. This combination of sizing equipment was discussed in the previous chapter in coextrusion, in the section on foam-cored coextruded DWV pipe.

The densities can range down to as low as 0.05 g/cc for foam tubing. The insulating characteristics are good with both polystyrene and polyethylene foams with low K values (K = thermal conductivity). One of the advantages of the polyethylene foam for this application is that it is flexible. If the tubing is slit while it is being extruded, the material can be wrapped around a pipe or duct for easy installation. The on-line slitter used would be similar to the one used in the tubular sheet extrusion process as illustrated in Fig. 8-13. Obviously only one slit would be made for the tubing application.

A wide variety of combination foam structures can be made using the foam-extrusion process in conjunction with solid extrusions. These are usually employed as structural skins to improve the load-bearing capability of the foam part. The pipe coextrusion was discussed previously. Another composite which is under development is foam-cored plastics siding where, in addition to adding stiffness to the siding material, the foam adds to the insulating value.

So far we have only briefly mentioned rigid PVC foams. In fact, they are widely used in the CBA process to make a wide variety of profiles used as moldings, pencil covers, picture frames, and even as a fencing material. Several tooling techniques can be used to make foam profiles. The extrusion die design used depends on the type of postdie forming fixtures used. Figure 8-16 illustrates the simplest approach. The die consists of a thin-walled smaller version of the profile to be made. The sizes are determined by the degree of expansion that will occur as the

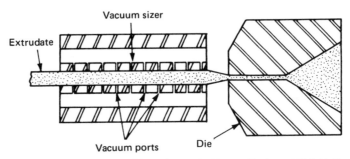

Reprinted from Plastics Machinery & Equipment 6(11), 11, Nov. (1977).©
Fig. 8-16. Foam profile extrusion die using free expansion and vacuum sizer.

plastic foams after leaving the die lips. The vacuum sizer unit has the dimensional shape desired and the expanding foam fills the sizer opening to form the part. This scheme has several drawbacks, especially if the shape is a complex reentrant form. Frequently the shape will jam in the sizer, and occasionally the material will fold and leave a deep line on the part.

Figure 8-17 illustrates an alternative die and sizer combination. In this case the die consists of a unit that will extrude a thin-wall shape that matches the actual outside dimensions of the shape to be made. The sizer unit is brought right up to the die opening spaced away with a thermal block that reduces heating of the sizer. As the material expands, it moves inward to fill out the shape. In order to expedite this, the center core of the die should be vented as shown. Among the advantages of this approach are the formation of dense skins on the part and the reduction in the tendency of the system to jam. It is also possible to apply a vacuum to the vent line which will aid the expansion process and can lead to lower foam densities for the same blowing agent concentration. An

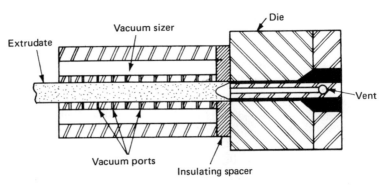

Reprinted from Plastics Machinery & Equipment 6(11), 11, Nov. (1977).©
Fig. 8-17. Foam profile extrusion die using inward expansion and vacuum sizer.

undesirable consequence of this method is that it can cause a hole or "pipe" running up the center of the extrusion.

Figure 8-18 illustrates the Celuka process tooling. In this system the material expands inwardly, as in the preceding example, but the expansion occurs in a controlled manner. This is done by the use of an internal mandrel which is shaped to permit controlled expansion of the material as it leaves the die. This eliminates the tendency for possible internal voiding and is claimed to produce finer quality foam. The most important part of the tool design is a knowledge of the proper mandrel shape to permit proper expansion. The shapes are based on empirically determined information with special test shapes. It is difficult to predict the mandrel shape based on materials rheology because of the complex expansion patterns.

Another patented process for making foam shapes relies on a stream separation scheme as illustrated in Fig. 8-19. Here an outer layer is separately passed through the die under conditions which make it high in density. The core material passes through a very short die which permits maximum expansion of the foam. The result is an extrusion with a dense skin and much lower density core without the use of a coextrusion procedure.

There is another technique used for postdie shaping of foam extrusions that is usually employed with CBA foams of moderate density made into thin-wall parts. The tooling is shown in Fig. 8-20. The material is extruded from a slit-type die and formed into a shape by the cooling blocks. The die slot is usually only 0.010 in. (0.254 mm) in height and the foam will expand to 0.050-0.060 in. (1.27-1.52 mm) in thickness. The die is a high-shear unit and accelerates the foaming process to give generally lower-density foams with the same concentration of CBA. The range of density is down to 0.25 g/cc, which is low for chemically blown foams.

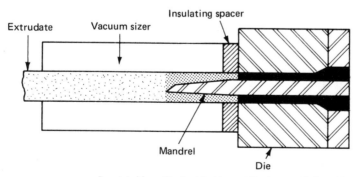

Reprinted from Plastics Machinery & Equipment **6**(11), 11, Nov. (1977).©

Fig. 8-18. Foam profile extrusion die using the Celuka process internal expansion mandrel.

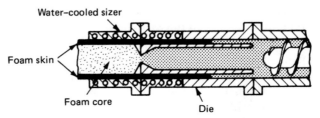

Reprinted from Plastics Machinery & Equipment **6**(11), 11, Nov. (1977).©

Fig. 8-19. Skinning foam profile extrusion system using twin stream technique (Reifenhauser).

The sizing and shaping units are water-cooled shaping blocks through which the foam strip passes after leaving the die. The water streams impinge on the outer surface of the extrudate, cooling it and lubricating the passage of the material through the cooling fixture. The surface quality is excellent and the surface quenching done gives the foam a dense skin and low density core. One of the significant advantages of this process is that the tools are low in cost as compared to most foam-

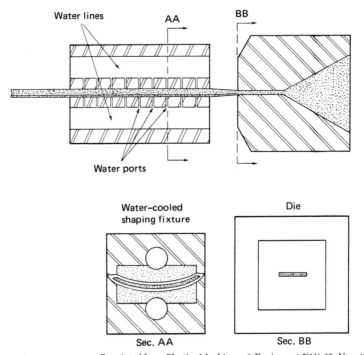

Reprinted from Plastics Machinery & Equipment **6**(11), 15, Nov. (1977).©

Fig. 8-20. Extrusion die and water block cooler sizer for curved thin wall foam slat.

extrusion dies and sizers. The slit die is simple to make and can be used for a number of shapes provided that the ratio of width to thickness is correct. The water-cooled slip blocks are easier to make than vacuum-sizer blocks and do not need the vacuum system and the vacuum ports. Operation is also easier since the blocks can be separated to thread up the line and then closed on the strip. The shaping block can be changed using the same strip to make several profiles. The only limitation on the technique is that it is applicable only to thin constant wall profiles which have none of the reentrant sections that would preclude closing the blocks after threading. There can be some variation in wall thickness, but it results from different densities.

Summary

The foam systems described are a few of the possible arrangements and materials that can be processed into cellular materials. The engineering plastics, such as polycarbonate, thermoplastic polyesters, and a whole range of other materials, can be extruded as foams. In practice, most of these extruded foams are made by some variation of the CBA systems. There has been some work done with solvent-blown systems on ABS and polyester resins. In addition to rigid materials, soft vinyl compounds and thermoplastic elastomers are extruded into foam using some variation of the methods described. Most products are closed cell materials, but changes in the technique and postextrusion processing can result in open cell foams.

The low cost and interesting properties of cellular plastics make them a significant group of materials for a wide range of extruded products ranging from packaging to gasketing, insulation, and structural foam sections. Further improvements will widen the uses for the materials by including fillers and reinforcements in some of the foams.

References

1. S. Levy, "Foam Profiles: New Processes, New Tools, New Products," *Plastics Machinery & Equipment* **6**(11), Nov. (1977).
2. S. Levy, "Extruding Plastics Foam Insulation," *Plastics Machinery & Equipment* **8**(1), Jan. (1979).
3. C. A. Villamizar and C. D. Han, "Studies on Structural Foam Processing I—The Rheology of Foam Extrusion," *Polymer Engineering & Science* **18**(9), July (1978).
4. S. Levy, "Solvent Blown Thermoplastic Foams," *Advances in Plastic Technology* **1**(1), Jan. (1981).
5. R. Ragazzini and R. Colombo, "Screw Extruder for Thermoplastic Foams," U. S. Patent 4,222,729, September 16, 1980.

Chapter Nine
Extrusion System Design and Integration

The success of an extrusion operation depends on the integrated operation of a number of pieces of equipment under proper control. It is necessary to determine the sizes of equipment required and the operating characteristics in order to design the overall system within the production and quality limits required. This chapter will cover the requirements for equipment and controls to give economical production with minimum capital investment costs.

One of the first steps in the line system design is selection of the equipment. Figure 9-1 is a master line which includes a number of different pieces of equipment for some stages of the process such as in-line static mixers, various cooling and holding systems for the extrudate, different puller options, winder and cut-off options, as well as control points for the system. It is useful to consider the master line in order to determine how each of the functions will be done on a line that is being designed for a new product.

The equipment is selected on several bases. One consideration is the capacity of the units in terms of product size and production rates. Another consideration is the effect of the performance of a specific unit on the product quality with respect to dimensions and physical properties. Still another consideration is the personnel and supervision required for the operation of a specific unit. An automated line may not be justified when the product is run only for a shift to make a minimum order. On the other hand, a product made in large quantities on multiple lines will justify sophisticated automatic handling and operating control to minimize the amount of labor input to the operation. Other equipment selection criteria include ease of start up, stability of operation, flexibility in handling a range of product variations, and low maintenance levels. A consideration in selection is the availability of personnel to operate sophisticated control equipment.

A discussion of system design is difficult without the use of specific examples. Many of the selection decisions are product oriented and are based on the production requirements. Some specific lines will be discussed to show how the equipment selections are made and how the equipment is sized to get the maximum performance needed. Since the state of the art in systems for different products has reached different degrees of sophistication, it will be evident in covering some of the systems that techniques used on one line could be used on other lines not now utilizing the technique, to get improved performance.

Tubing extrusion systems have a wide range of options that make

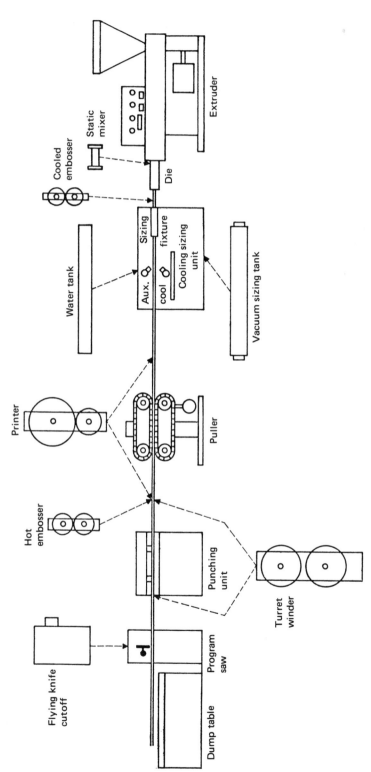

Fig. 9-1. Extrusion system master line diagram. Many of the options for making a specific product are shown. Alternatives are selected on the basis of the product requirements.

them one of the most interesting to examine. In addition, there is available a wide variety of sophisticated controls that are used on some lines to keep the quality and dimensions within close tolerances and also substantially improve the line performance. The elements in an extrusion system for tubing are

1. Extruder
2. Dies
3. Sizing equipment
4. Puller
5. Cut-off unit
6. Alternative winder

There are also a number of optional units used on tubing lines:

7. Static die-head mixer
8. Extruder valve
9. Cooling tanks
10. Vacuum sizing tanks
11. Water-cooling troughs
12. Dry vacuum sizers
13. In-line marking units
14. Internal mandrel sizing units
15. Diameter-gaging equipment
16. Wall-thickness gaging equipment
17. Dump tables and stackers
18. Hopper drying equipment
19. Acoustical flaw detectors
20. Other special equipment

Units that can be used on pipe and tubing lines far exceed the list given. The specific requirements of the product will generally determine the selection and specifications of the equipment to be used. The main selection, of course, is the extruder, which will be needed to supply the necessary amount of properly plasticated melt with a minimum in delivery variation. General sizing can be done from a table such as Table 9-1 which lists the nominal output of a series of extruders of different screw sizes and L/D ratios. These rates are given with the typical

Table 9-1. Typical Production Rates for Single-Screw Extruders

1½ in.	(3.81 cm)	24:1 L/D	50–60 lb/hr	(22.7–27.3 kg/hr)
2 in.	(5.08 cm)	24:1 L/D	90–120 lb/hr	(40.1–54.5 kg/hr)
2½ in.	(6.35 cm)	24:1 L/D	150–250 lb/hr	(68.2–113.6 kg/hr)
3½ in.	(8.89 cm)	24:1 L/D	300–400 lb/hr	(136.4–181.8 kg/hr)
3½ in.	(8.89 cm)	30:1 L/D	350–450 lb/hr	(159–204.5 kg/hr)
4½ in.	(11.43 cm)	24:1 L/D	700–1,000 lb/hr	(318.2–454.5 kg/hr)
6 in.	(15.24 cm)	24:1 L/D	1,200–1,600 lb/hr	(545.4–727.3 kg/hr)

polymers that can be extruded by the machine using the appropriate screw for the material. The procedure for selecting a screw involves trying several that have been run successfully with the same or similar materials on products with the same size range.

Among the other factors involved in selecting a suitable extruder are the motor horsepower, the variable-speed drive system, the choice between single- and multiple-screw machines, the temperature control system, and the other instrumentation on the machine. Other specifications necessary are vent and vacuum pump sizes on vented extruders.

The extruder choice is based on the material to be processed into pipe or tube. Single-screw machines are used for polyolefin materials, ABS, polystyrenes, and most of the commonly used thermoplastics. The twin-screw machines are used primarily for rigid PVC pipe, especially the larger sizes. The horsepower requirements depend on the resin processed and the output requirements. Stiff resins such as rigid PVC and polycarbonate require larger horsepower for the same screw sizes. Materials such as low-density polyethylene require much lower power for the same extruder output. The extruder performs the functions of heating the resin, plasticating the resins, building pressure in the resin melt, mixing the melt to obtain a uniform material for the output, and feeding the material through the die.

The selection of the proper machine is determined by the needs of the plastic that is to be processed. The amount of heat input needed for the required production rate is one factor that must be considered. Depending on the characteristics of the material, this heat can be supplied by shear heating of the material, or by heat transferred through the extruder barrel into the melt. The machine choice will be affected by the requirements for shear and the requirements for heat transfer to and from the melt. If lower and controlled shear is needed as, for example, to extrude rigid PVC, either a twin-screw machine or a single-screw machine with a low-shear screw will be chosen. Heat effects are also influenced by the hopper drying unit. If the material enters the extruder at a higher temperature than ambient, it reduces the heat transfer needed in the machine; it can also affect the amount of shear heat generated by the machine and the degree of mixing that occurs in the extruder.

In most cases there are several machines that can be used for a specific resin and product. The performance of several combinations of machine, hopper drier heater, static mixer (if feasible), and die resistance should be examined to compare the expected performance against the system to determine which combination requires the least capital cost per unit of output. In addition, the energy and services costs for the several alternatives should be evaluated to determine which combination will have the lowest operating costs. Another factor that will enter into the selection is the melt quality and melt quality stability of each of the alternative systems. This is a matter of experience, and the machinery manufacturers can supply some guidance for the choice. In other cases it may be desirable to consult with an engineering organization that has had experience with the resin to be used. Frequently, the determining factor in selecting an extrusion machine is the melt quality and output stability.

The dies are the major shaping element in the pipe or tubing lines. Their design and construction have been covered in the earlier discussion. Again, which type of die is selected for the system depends on the material that is processed. It also depends on the dimensions of the product, the production rate, and the lengths of the runs that will be made. If the materials used are stable at the extrusion temperatures, the dies can be fairly simple in design without undue attention being paid to material hanging up in the die in stagnant-flow regions. On the other hand, if the material tends to degrade at the extrusion temperatures it becomes important for long runs that the dies be completely streamlined without any stagnation zones in the tool that can lead to the production of streaked and burned product.

The land lengths and the shapes of the approach angles in the dies have been discussed in the prior sections of the book. Each material has characteristic requirements for these parameters. The choice of design, as far as the system is concerned, is dependent on the degree of streamlining needed for the production schedules used, as well as on the degree of maintenance each design will need in the long run. Some die designs will exert different back pressures on the extruder and thus will affect the machine performance. Higher back pressures usually lead to higher shear levels with additional shear heat generation. Lower back pressures may lead to undesirable output fluctuations in output and melt quality. In Ref. 1 the marrying of the die to the machine is discussed. This is another consideration in the system design.

The simplest die constructions are used with dies for low-density polyethylene. Since there is little corrosion potential, the dies do not need plating maintenance. The streamlining requirements are not severe, and even if there is some hangup in the die, it will not lead to serious changes in the product quality. The most difficult dies to design and produce are those for a heat-sensitive material, such as rigid PVC, especially when they are to be used to produce pipe on a continuous basis. Careful attention to streamlining is essential, and the dies must be made to minimize any high-shear sections. The tools must be plated to prevent surface degradation of the steels in the dies, and the platings must be renewed on a careful schedule to prevent poor product or low production rates.

Die selection must also take into account the effects of the internal structure of the tool on such problems as welds caused by rejoining of the melt stream after passing around a mandrel or over spider legs in the die. The design selected should produce a negligible weld line in the tubing or pipe at the maximum production rates needed. As in the case of the extruder package, experience is the best present guide to selection of the die design for a material and size range of tubing.

Cooling and sizing equipment selection for the system is dictated by the same considerations as the other line components—rate, material characteristics, dimensional control, and product size—but the parameters are different. The cooling system can be a simple water tank, a vacuum sizer, an internal cooling mandrel, an air-cooling system, or a combination of several of these. Since the tubing is sized as it is cooled, the sizing devices are intimately integrated into the cooling unit.

One of the simplest approaches is to pull the tubing directly into a water tank. This method is frequently dictated by the materials surface friction characteristics when hot and by the stiffness of the softened plastic. Flexible polyvinyl chloride formulations have very tacky high friction characteristics and are very stretchy materials. As a consequence, the only practical way to cool these materials is by taking them directly into a water tank. The sizing is accomplished by control of the relative pulling rate to extruder delivery rate. There is usually little or no fixturing in the tank because of the friction problems. Occasionally, water-cooled rings will be used to prevent flattening of the tube as it goes through the tank. Controlled sizing is not used in this arrangement.

One of the most widely used schemes in tubing and pipe systems is the use of a vacuum sizing tank. The outside diameter of the tube is set by holding it against a set of sizing rings which are immersed under water in the vacuum sizing tank. The water in the tank is held at less than atmospheric pressure so that the tubing is held against the sizing rings, while the outer layers are set by contact with the cooling water in the vacuum tank. This method is very effective with medium- to thick-wall tubing in the range of 0.030–0.500-in. (0.76–12.7-mm) wall. The vacuum section of the cooling sizing unit will set the entire thickness at the low end of the range. Additional cooling tanks and evaporative cooling are required at the upper end of the range. There are two systems problems in the vacuum sizing method. One is that the amount of cooling that can be done by the unit determines the length of the sizing unit. The other is that drag effects are produced by intimate contact with the sizing rings. The contact pressure is determined by the vacuum level in the tank and this, in turn, determines the amount of drag that will occur. The longer the vacuum-cooling section, the larger the surface area for cooling and the greater the drag. Selection of the sizer will be based on the best compromise between these factors and the additional cooling needed will be done in atmospheric pressure units downstream of the vacuum sizing tank. It is apparent that the drag effects are important with respect to the puller.

The other methods of sizing and cooling are by the use of an internally cooled mandrel (commonly used for thin-wall tubing), the use of dry vacuum sizing units, and the use of internally pressurized sizers. The internal mandrel system is used to make polyethylene tubing with walls of 0.006 to 0.020 in. (0.15 to 0.51 mm) and diameters of ½ to 3 in. (1.27 to 7.62 cm). Dry sizers are used with materials such as the acrylics that cannot be rapidly cooled because of a tendency to internal void formation in thick-wall sections. The internal pressurization system is used for very-heavy-wall pipe with wall thicknesses in the 0.250–2.000-in. (0.64–5.08 cm) range. The vacuum system does not give effective contact with the sizing rings when extruding stiff materials such as rigid PVC when the walls are very heavy.

The pulling unit has an important function in the system and is usually not investigated as carefully as it should be for performance characteristics. In the free-extrusion sizing system the puller speed relationship to the extruder output is the controlling factor in the size of the product. In the vacuum sizing cooling arrangement the outside diameter is controlled by the sizing sleeves, but the wall thickness control is dependent on the puller speed control. The factors which determine the puller operation effectiveness include the power capability of the unit and the grabbing effectiveness of the puller belts or rollers. A puller to be used to pull pipe or tubing through a long vacuum sizer needs a higher horsepower drive than one used for free extrusion. Another system tradeoff that is required is the comparative size of the puller and puller drive versus the length of vacuum-cooling surface in the sizer.

The belt surfacing of the puller is important to the control of the product size and the uniformity of the product. It is necessary to design the surfaces and select the surface materials to minimize slippage of the tubing in the puller. Constant slippage will tend to make the product larger than the calculated size. Intermittent slippage or slip-stick phenomena will cause variations in the dimensions of the product. The use of belt surfaces that have high coefficients of friction against the plastic being extruded is important. In addition, the belts or puller rolls should be contoured to increase the amount of friction surface available. Careful control of the pressure of the pulling surfaces on the product is necessary since excessive pressure will crush the product and inadequate pressure will increase slippage in the machine.

Puller-speed control systems which will overcome the effects of slippage on the product should be used. The speed of the extrudate should be measured directly and the belt's speeds should not be the speed that is read for control. A number of different control units have been devised to do this, and they should be used whenever the dimensional control is critical.

The speed control on the puller is important in the cut-off operation that cuts the product to length. In the case of the cut-off systems that predetermine a length by measuring the amount of material that has moved by, the length is counted on the belt of the puller. This is inadvisable for precision length control if there is any appreciable slippage in the puller. In the synchronized cut-off saws which use a limit switch to initiate the cut-off, smooth control of the extrudate speed is also necessary to prevent the saw from binding against the material as the cut is made. Here again, if there is any appreciable slippage, it is advisable to use an independent line-rate sensor to operate the saw cut-off drive.

The puller constancy is also a factor in the collection of product by winding, although it is less important than in the cut-off collection arrangement. The winders are usually constant line-speed units that will tend to produce excessive slack or excessive tightness if the line speed varies. Since these units are usually equipped with a line-tension adjustment, such as a dancer roll which will compensate for the line-speed variations, it is not as significant a factor for winding as it is for the cut-off operation.

Other units in the system requiring synchronization with the line speed would be marking or printing units and in-line fabrication punches which might be used to make cuts or holes in the tubing. The other line components that must be considered are the control units used to maintain product quality and dimensions.

The control equipment for the system includes the temperature, speed, and pressure controls on the extruder; the temperature and circulation rate controls on the cooling units; and the puller line-speed controls and associated devices for length cut-off control. The other associated instrumentation would include dimensional gaging units to measure the outside diameter and wall thickness of the product as well as the variation in wall thickness and eccentricity in the tubing. In addition, an ultrasonic flaw detector may be used for determining the quality of the material in the tubing walls. Many lines do not have all of this equipment, and the choices will depend upon the degree to which the process needs to be controlled to make acceptable product. In some systems the gaging equipment is coupled into a closed-loop control system to automatically adjust the operating conditions to maintain dimensions within limits.

The ability to introduce a closed-loop control into the puller portion of the line to maintain dimension is based on having a puller that can have its speed changed by an error signal generated by the gaging equipment. SCR drives and variable-speed AC drives are suitable for this. The other factor that must be considered in the closed-loop control is the sizing methods used. In the case of a vacuum sizer or an internal mandrel sizer, one of the dimensions is an invariant determined by the gaging section of the sizer. Adjusting the puller rate will then affect only the wall thickness. For the outside sizing the OD can be used as the major control and the ID can be controlled by the rate changes. For ID sizing the ID must be gaged with the wall thickness. This is more difficult with present gaging sensors than OD gaging. The free-extrusion method used with the soft PVC formulations can control only one dimension. The other dimensions will fluctuate to reflect all of the process variations. Stability in extruder output is particularly important for good control of the product for this operation.

This review of the pipe-and-tube system gives a basic insight into the interaction of the various parts of an extruder system and can be the basis for a systems approach used to design any system. The system in a functional block diagram is shown in Fig. 9-2. The interrelationship between the system elements and the control units is made clearer by examining the connections between the parts. The extruder has its own integrated control arrangement designed to give stable operation. The control package shown can vary from the conventional manual-set discrete controller setup to an integrated control system which maintains the interaction desired between, on the one hand, the desired output rate, melt temperature, and melt pressure and, on the other hand, the controller settings to a computer-controlled closed-loop system which will compensate for material and other variables to give consistent output rate and condition (this will be discussed in a separate section on computer control). Whatever the package, the output from the machine is fed to the downstream units. The dimension sensors can be used to record and indicate the part sizes or the data can be used to generate an error signal to the puller SCR control, which will then vary the puller speed to correct the size to the desired range. This can be done open loop or closed loop and, again, the control package can be equipped with a microcomputer that will take into account the variables that affect the size and make the corrections in an anticipatory manner. This system will require the two-dimensional sensors shown—one at the die and one past the puller.

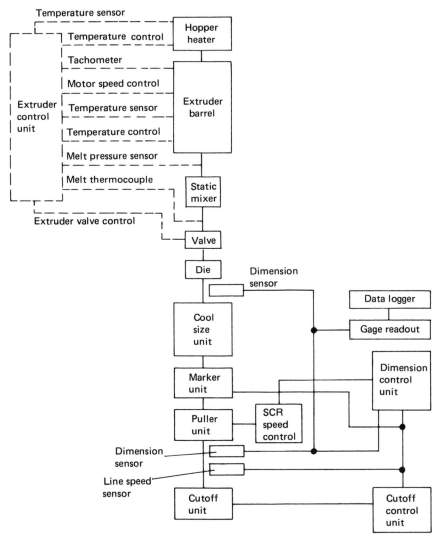

Fig. 9-2. Systems diagram for tubing and pipe extrusion. The interrelationships between systems elements are diagrammed to show interaction.

The line-speed sensor is needed to properly operate the cutoff unit to control the piece length. The same signal is fed to the marker unit so that its speed can be synchronized to the extrudate speed rate to give good marking. There are no specific controls shown on the sizer cooler unit for vacuum level and coolant rate conditions, since these would be made a part of the unit and, in general, do not have to be interactively integrated to the other elements in the system.

From the systems diagram it is apparent that there are a large number of interactions between the various parts of an extrusion system. The diagram does not cover the extruder downstream interaction which is frequently included in a computer-controlled-system concept. This complete control has not yet been introduced into the pipe-and-tubing production lines although it has been on sheet and film lines.

The same type of analysis can be applied to the other production systems that are used in extrusion. Sheet, film, profile, covering, and the other product lines that have been discussed would use the appropriate extruder package with the necessary downstream units to make the product. Sizing of the units and their selection would be based on rates of production required and the utility of a particular piece of equipment for the product size and characteristics needed. The same resin considerations are needed for the other products as for the one in the pipe/tube illustration.

The approach is to list the functions the line must perform. From this list a systems diagram such as the one in Fig. 9-2 should be constructed in order to determine what units are required. Additional information, such as the line rates of speed that correspond to the weight per unit time of the extruder output, will permit selection of units with the proper speed ranges needed for pulling, marking, and other operations whose performance is related to the line speed. Dimension gaging functions are added to the system considerations to control the product size and to minimize variations. Depending on the criticality of control required, open- or closed-loop controls can be used and the process placed under computer control if justified.

Examination of a sheet line from the systems standpoint is informative with respect to the control and component selection, and the interactions between the various elements of the system. This is also useful because sheet lines have been built with the most sophisticated on-line controls used in any extrusion operation. Figure 9-3 is a systems diagram for a sheet line using a die whose opening can be adjusted during the operation of the system. Also indicated in the diagram are the additions required to run coextruded sheet materials.

The line components include
    1. Extruder package
    2. Sheet die with Autoflex die-control lips
    3. Three-roll stack
    4. Thickness sensor system with cross-sheet traverse
    5. Automatic feedback control system for sheet thickness control
    6. Automatic sheet cut-off units
    7. Sheet stacking unit

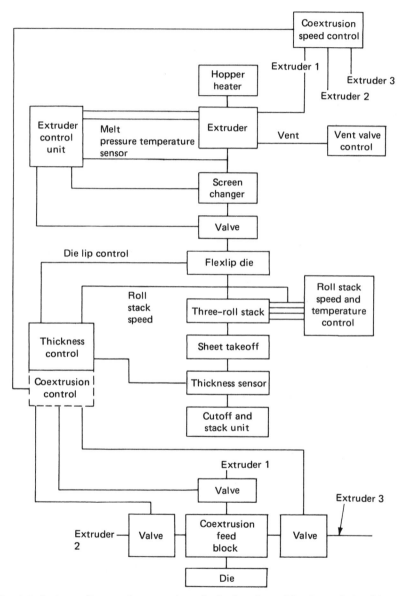

Fig. 9-3. Systems diagram for extrusion of plastics sheet. The interrelationships and interactions are shown for both single-layer sheet and coextruded sheet.

The extruder package is designed to deliver the required line output of material for the basic sheet line. These machines will generally deliver at peak capacity for this type of operation. For a number of materials made into sheet (such as ABS resin, high-impact polystyrene materials, and acrylic resins), a vented machine is desirable to give the best extrudate quality. This feature is shown in the systems diagram. The new portion of the extruder package requires additional instrumentation and controls to properly operate the vent to remove volatiles from the melt without overflowing the vent or starving the second stage of the extruder. This is generally a manual operation—setting the back pressure on the first stage of the screw to produce complete resin plastication, and doing this at a throughput to match the pumping capacity of the second stage in the screw. In some instances the restriction is an adjustable valve and it can be controlled with a closed-loop control unit.

Stable output from the extruder is necessary, and this is accomplished either by careful settings on the machine, plus the use of closely controlled feed materials, or by a feedback control system on the extruder. The sheet-line machines are the ones that have been operated successfully with an on-line closed-loop computer-control system which will be discussed in detail in a later section. Variations in output from the extruder produce a control load on the system that is used to control the gage of the material.

As can be seen from the systems diagram (Fig. 9-3) the interactions of the various elements of the system can be rather complex. The extruder-control system in the single-layer mode will deliver a constant flow of material to the die. The automatic screen changer will filter any debris from the melt, and its operation timing will depend on the sensed pressure drop across the changer unit. It is necessary to do this to avoid a slow buildup of back pressure in the screens.

The Flex-lip die and the roll stack interact in a complicated manner to control the sheet thickness. The die lips and the opening in the roll stack are set to the desired range. The roll-stack speed is set in the appropriate range to have the surface speed of the rolls match the speed of the sheet coming from the die. If there are variations in thickness in the machine direction sensed by the thickness sensor, two things can be controlled. One is the speed of the roll-stack drive and the other is the die-lip opening. A control strategy must be developed to determine which of these is adjusted and to what extent to correct the thickness variations. Generally, long-term changes in thickness involving several minute time constants are adjusted by speed changes. Short-term fluctuations are adjusted by operating the die lips. The roll-stack opening is an invariant quantity in these adjustments and is usually not changed during operation.

The cut-off units are constructed with either a saw or shear to operate in a manner similar to the one shown for pipe. It involves length sensors and synchronized drives for saws, and the general comments made about the units are the same. Usually sheet is not embossed on line, but in the cases where it is, the same synchronizing requirements are necessary as for the marking units used on the pipe lines.

Introducing two additional streams in a coextruded sheet line system adds substantially to the complexity of the operation. The relative flow of the materials from the multiple extruders must be controlled. In the alternative coextrusion setup shown in the systems diagram, two flow-control strategies are indicated. One is the use of extruder speed control, and the other is the use of valves operated from information input from the thickness sensor system. This unit is now more complicated since it must measure the thickness of several layers of material as well as the overall sheet thickness. For the sake of clarity not all of the interrelated controls are shown in the system. It is easy to see why the sheet-extrusion lines were the first to be considered for a process computer control. This is especially true for the coextrusion systems where the interaction of the separate feed systems is involved. With a complete data evaluation the operation is much easier to start and to keep under control.

Equipment selection for the line is dictated by the system considerations. The die is one with the Autoflex feature for on-line thickness control. The die-setting range will be determined by the product required. The screen changer is selected to match the machine throughput and the mounting requirements. The valves used will be adapted to the degree and method of control needed. They can be adjusted automatically or adjusted manually to match the required pressure and flow needs. The three-roll stack is selected on the width and thickness requirements of the product. Since the cooling capacity of the rolls is usually the limiting factor in the line rate, the roll size (diameter) is made accordingly and the cooling system is designed for maximum cooling. The speed control on the stack is usually an SCR control or similar unit whose speed can be adjusted with a reference signal to suit dimensional control requirements. The control unit which controls the Autoflex die uses input from a traversing nuclear gage sensor. This must be augmented by other sensors such as ultrasonic units for the layer sensing for the coextruded form of product.

The examination of the sheet line from a systems viewpoint clearly shows the specifications needed for the line components and controls and is very effective in sizing the units and determining speed ranges for the drives, line rates for the product, and other operating parameters for the system. It is also useful for finding critical control requirements for the product.

Summary

This section has been concerned with the systems design aspects of extrusion lines. The same approach that was used to analyze the pipe/ tube system and the sheet lines system can be used in analyzing wire systems, foam systems, profile lines, covering lines, and other systems for extrusion. The approach is to define the functions required to process the material into the product with the requisite characteristics and quality level. In each case these would include the extruder as a plasticating device for the material with such modifications as are necessary to provide melt at the desired quality level, pressure, and temperature to make the product at the desired rate. The other components of the line will shape the extrudate, set it, control its shape and dimensions, collect it, and provide means for control of the line and of the extrudate quality.

Reference

1. S. Levy, "Extrusion Systems—Putting It All Together," *Plastics Machinery & Equipment* **9**(5), May (1980).

Chapter Ten
On-Line and Computer Control of the Extrusion Process

Continuous processes, such as extrusion, should lend themselves to systems of continuous process control, like the ones used in other chemical unit operations. Polymer materials are somewhat more complex than the usual fluids processed, and the historical development of the extrusion process was such that the typical control systems approach was not considered. With the development of the equipment with better-defined process actions on the materials, better understanding of the rheology of plastics melts, better data on the materials, and, most importantly, much more critical requirements for products, there has been a great deal of activity on the development of close control systems for extrusion.

Extruder Control

There have been a large number of approaches to the process control improvement. One parameter that has been subjected to study is temperature and temperature control. From simple on–off controls the machine sensing and control systems have been brought to the stage of read-around temperature measurements which control power input units that deliver the needed heating to promote a stable melt condition in the machine. The most sophisticated of these systems also measures melt temperature conditions at one or more locations in the melt stream and determines the interactive effects of shear rate and temperature to produce closely controlled melt temperatures.

Another control parameter that has been under study is melt pressure at the extruder head end. Improved sensing systems have been developed, and there has been significant information generated on the effects of flow restrictors such as screens and valves and on the effect of machine screw speeds on the head pressures. One of the problems is that the melt pressure is a result of several different factors, such as the temperature, amount of shear history, and the degree of mixing in the machine, so that it is difficult to control directly.

The important question to ask is, what is desired in the way of output from the extruder? Another way of stating this is, what are the requirements on the melt parameters to produce stable output from the extrusion die? It is necessary that the melt delivered to the die have the same pressure, temperature, and shear history at all times in order for the die delivery to be constant and stable. Since the shear history is a function dependent on the back pressure in the machine and this, in turn, interacts with the melt temperature, it is evident that the melt parameters are interdependent and strongly dependent on the machine settings and how it is operated. Present control practice takes these interactions into consideration only by means of the skill and experience of the machine operators and setup men. It would be invaluable in the operation of complex extrusion systems if the start-up and running strategies of the operators could be formalized and the operation be consistent in performance without relying so heavily on operator skills.

There are two obvious ways to control the output rate of material from an extruder. One is to change the screw speed. The other is to change the back pressure at the head end of the machine. These apply to all extruder types, but the main interest is the single-screw machine because its output stability is more difficult to control and it constitutes the majority of the systems in use. Each of these methods of control has effects other than simple delivery rate changes. Given a specific heat-transfer situation on the machine barrel, both will affect melt quality, especially with respect to shear history of the material. From the prior material on melt rheology it is known that increased shear will reduce the melt viscosity. This will, in turn, alter the machine delivery rate in a manner that is specific to a particular formulation of resin.

Changing the machine screw speed will increase the amount of material taken into the machine and will give increased throughput. It will also move the melting and plasticating regions in the barrel from the positions that existed before the speed change, and this will alter the heat-transfer conditions in the machine and the melting pattern. In addition, higher screw speed will result in additional shear on the material that will produce lower melt viscosity. This can result in more material slippage in the machine and the general effect is to change the throughput in a nonlinear relationship to the changes in screw speed. In addition, the shifting of the melt bed configuration can lead to a condition where the melt is not completely plasticated by the time it enters the metering section of the screw. The result of this is to permit unplasticated granules to be delivered by the machine and to initiate the condition called surging (wide swings in delivery rates caused by unstable flow).

Changing the restriction to flow at the entry to the die will alter output from the extruder by changing the back pressure on the machine. This can be done with a valve or similar means. The changed back pressure will also have an effect on the melt quality by changing the amount of shear and, thereby, the shear history of the material. There will be some slight effect on the melting bed profile in the machine but it is not substantial. The main point to recognize is that increasing the restriction will produce lower melt viscosity resulting from higher shear, and will result in the situation where the pressure drop will not be linear with the degree of flow restriction.

The nonlinear and interactive response of the flow rate to the controlling element makes closed-loop control of extruder output difficult. A specific control strategy must be devised to enable the controlling devices to eliminate changes in output. This is usually done with a microcomputer control unit which has data stored to indicate the performance of the machine with the resin being processed under changing machine conditions. The microcomputer will take data from the machine, the melt, and the machine resin input and compare this information with the stored data to determine the adjustments that must be made in either pressure or screw speed to maintain constant output rate.

The microcomputer can also make temperature adjustments on the machine barrel to compensate for the effects the changed machine conditions will have on the melt temperature. By using these inputs the machine output can be stabilized. Changes in incoming material temperatures will be corrected for before they affect output rate. The machine can also be made to compensate for changes in resin characteristics as the batches change, by storing the melt characteristics of each batch in the computer memory and indicating to the computer when the batch change occurs.

Figure 10-1 from Ref. 1 is a process block diagram of the extrusion process which indicates the factors that affect machine output rate and melt quality. This type of process is controlled by means of a closed-loop feedback control using the computer process control system shown

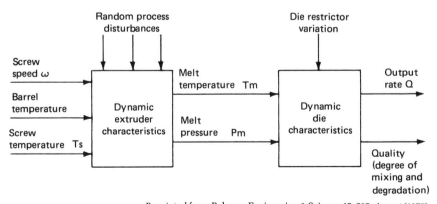

Reprinted from Polymer Engineering & Science 15, 595, August (1975).
Fig. 10-1. Extrusion process block diagram. Variables affecting the extrusion process are shown and the methods of control indicated.

schematically in Fig. 10-2. The reference discusses several computer strategies that can be used to control the output of the machine so that it produces a constant delivery rate of material with consistent melt quality characteristics. The study presumes an adequate knowledge of the operating characteristics of the machine and material so that a basic start-up sequence is determined as information that the computer has in its memory. The primary function of the machine is to maintain the performance of the unit.

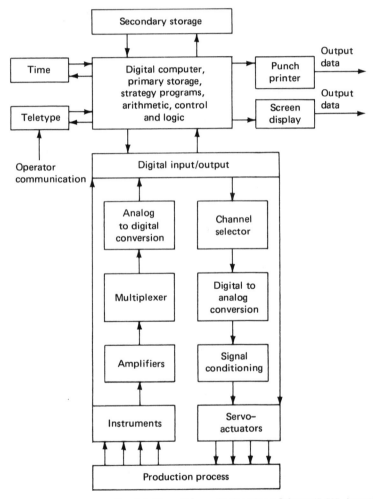

Reprinted from Polymer Engineering & Science 15, 595, August (1975).
Fig. 10-2. Process control system for plastics extrusion with process under computer control.

Since this paper (Ref. 1) was written, there have been several systems of adaptive control suggested and tried. These machines will start with a set of operating conditions that produce a reasonable level of stable output and then make variations in the machine settings to increase the output rate. The main method of determining if the adjustments are not destabilizing the machine is by sensing of fluctuations in output pressure which signal the onset of melt instability and surging. Another method is to make measurements of the melt quality by the use of sonic measurements on the exiting melt. Sonic spectra will indicate changes in shear history. Such changes will also signal the onset of surging or other unstable output conditions. Another change of direction in the control strategy is to put more emphasis on valving or flow restriction as compared with screw speed control for short-term adjustments of output (time constants between 0.1 and 30 sec), because of the availability of electroservovalves which can respond rapidly to signals.

The adaptive control strategy suggested above permits the system to improve its performance during normal operation. The changed operating conditions that are generated can be stored in the machine's temporary memory for reuse during the same run or can be removed from the temporary memory and stored on a fixed read-only memory (ROM) to be used later on the same system or transferred to another similar system to be used as the stored operating conditions for the system.

The stored data needed to make the extruder computer control system operative are rheological data on the resin being processed and a model showing the manner of operation of the machine in response to the changes in the control functions. In addition, a start-up and run set of instructions are needed to bring the system into operation. These are generated by an operator with experience and knowledge of the process.

The means of effectuating the control is by changing the machine operating parameters. To change screw speed in response to information from the control computer, the main drive motor should be equipped with a signal-sensitive drive, such as an SCR drive. To change the restriction to flow at the extruder head, a signal-responsive melt valve must be used to adjust the restriction to that dictated by the computer. The temperature control system should be capable of adjusting the machine temperatures in accordance with signals from the control computer and temperature sensors to adjust melt temperatures. The melt temperature will vary as a result of the rate adjustments and variations in the character of the incoming feed. The computer can also be made to compensate for variations in the environment surrounding the extruder.

The key to a successful closed-loop control system for extrusion is stable output of the machine through the die. The remainder of the system can be controlled by other schemes which control takeoff rate and other downstream parameters.

Downstream Control

For process control considerations the downstream units to be considered are

1. The die, especially if it is adjustable
2. Roll stacks, vacuum sizers, or other units which set one or more dimensions on the extrudate
3. The pulling system

Assuming that the extruder output is stable, the other determining factors in dimensional control are related to the rate at which the product is drawn away from the die. This is the pulling function. The puller units, whatever type is used, can have their speed varied to adjust the dimensions of the product. In order to do this with closed-loop control systems, it is necessary that the units be equipped with drive-control systems which can vary the drive speed in conformance with a control signal, such as an SCR drive.

As was previously mentioned, pulling systems are generally not positive in their grip on the extrudate. As a result, it is important in the control consideration to sense the line speed with an independent speed sensor to be able to effectively control the pulling rate. Several such sensors are described in Ref. 2. One of the best is a lightweight rolling contact wheel. This information is needed in conjunction with dimensional sensors to provide the information for closed-loop control.

Dimensions are sensed in a number of different ways—ranging from mechanical contact through nuclear gages to optical sensors using lasers. Nuclear gages are common for sheet lines. For tubing, sonic devices and optical sensors are used to measure size and wall thickness. Table 10-1 is a table of the various sensors. The location of the sensors is also important. In order to get effective response times at least one of the sensors should be located near the die so that the changes in dimension are sensed as soon as the product leaves the die lips. Reference 3 discusses the requirements of the gaging in detail. Information signals from the dimensional-sensing arrangement are processed so that changes in the puller equipment speed are made to correct for dimensional shifts. This can be a fairly simple feedback system, but it must be designed to avoid an overreaction which can destabilize the control. One method used is to limit the control range to a fraction of the speed range of the SCR or other controller. If the deviation gets outside the range, an

Table 10-1. Restrictions on Various Gaging Methods

	Ease of Application	Accuracy	Dimensional Range	Transverse Measurement	Reflex Measurement	Freedom From Interference	Material Dependency	Sensor-size Limitations	Complexity of Equipment	Ease of Calibration	Cost	Sensitivity
Rolling-contact	Easy	Good	Wide	Yes	No	Good	No	Low	Low	Easy	Low	Med.
Air	Fair	Good	Wide	Yes	No	Good	No	Some	Med.	Easy	Med.	High
Magnetic-reluctance	Easy	Fair	To ¼ in. (0.64 cm)	Yes	Possible	Fair	Some	Some	High	Easy	Med.	Med.
Sonic	Fair	Good	To 1 in. (2.54 cm)	Yes	Yes	Fair	Yes	Some	High	Fair	Med.	Med.
Optical	Fair	Fair	Wide	Yes	Yes	Good	Some	High	Med.	Fair	Med.	Med.
Laser-intercept	Easy	Good	Wide	Yes	Yes	Good	Some	High	Med.	Fair	Med.	High
Laser-interferometry	Hard	Exc.	Ltd.	No	Yes	Good	Yes	Some	High	Easy	High	High
Capacitance	Easy	Good	Med.	Yes	No	Fair	Yes	Low	High	Easy	Low	Med.
Proximity	Fair	Good	Wide	Yes	No	Fair	Some	Low	High	Easy	Med.	High
Beta-ray	Fair	Good	Ltd.	No	Yes	Good	Some	Low	High	Easy	High	High

alarm sounds to alert the operator and manual correction can be made to the line speed.

The problem is complicated when the puller system has slippage, especially random slippage, during its operation. This slippage must also be corrected for, and there must be a method for mixing the two dimension-affecting signals so that the speed adjustment is correct. Frequently, it is necessary to again go to a microcomputer system to handle the data in order to produce the necessary process control. This is especially true when the system uses another sizing unit, such as the blow-up unit in blown-film extrusion, or sizers and roll stacks in the sheet system.

We will consider the blown-film problem later, but it is useful to inspect the sizer situation and the roll stack arrangement to see how these can affect the control system requirements.

The use of a sizer introduces an invariant into the takeoff system. On tubing, the sizers have the effect of fixing one dimension, usually the outside diameter, by means of the sizer. As a result there is one primary dimension, namely, the wall thickness, to be controlled by the control unit. In addition, the control system may deliver information for correction of wall eccentricity or, conceivably, control the wall eccentricity. In that case the puller information will be derived from the wall thickness measurement. By processing the mean wall thickness gaging and comparing it with the desired dimension, the puller can be speeded up or slowed down to provide constant wall thickness. Eccentricity information can be fed back to a die adjuster to move the pin or bushing to make the inner and outer wall concentric.

The die adjustment situation is more significant in sheet lines since adjustable lip dies are widely used. Autoflex dies will change their opening in response to signal information from a gaging system used on the sheet and will do this locally across the die as well as completely across the die. A problem exists because the roll stack is supposed to be an invariant thickness dimension control device, and it is usually not made responsive to the thickness variation signals. In practice it is not a total invariant in that thicker and thinner materials than the roll settings can pass through. It does, however, represent a complication in the closed-loop control. This is especially true because the sensors are usually located after the roll stack and do not exactly reflect the thickness profile at the die.

From this discussion it appears that the controls needed for both the sheet and tubing could benefit from the use of a computer-control system that processes the data on the dimensions and operation of the sizer equipment to control the puller speed. In the case of the sheet, the computer can make decisions as to which should be adjusted for machine direction dimensional—the die opening or the line speed—and, based on the time constant of the variation, adjust one or the other. In addition, the computer could permit an adaptive control mode. By recording data on the variation patterns of thickness a predictive equation can be generated based on rates of change. When a particular pattern of dimensional change appears, the control computer can anticipate the dimensional shifts and make compensations before the product approaches the control limit on size. A control computer can justify its cost by eliminating off-grade material, as well as by restricting the range of variation, and increasing the amount of shippable product per pound of output.

System Integration

From the discussion to this point it can be seen that both the extruder and the takeoff line can be independently controlled using simplified feedback or computer process control. The introduction of a variable die structure into a system can provide another means for controlling the product dimensions. It also introduces the need for integration of the two control systems. The die can be controlled either by the line inputs or by the machine requirements. We are introducing a regulatable entity into the system which can be used to control either output or line rate. In this instance, a decision is necessary as to which is the better mode of operation for a specific case, and in order to do this the two control systems must be integrated. This is done with the goal of maximum output possible of product with the necessary dimensional and quality controls.

The decisions as to which way to control are generally based on the time constants of the variations. Very-short-term variations, such as are produced by melt fracture, cannot be controlled by either control loop. Variations in the range of 0.5–15 sec are best handled by the takeoff control system since the response times are shorter. The exception is when there is an adjustable die. In that case it may be most feasible to use the die adjustment as the proper control element. In some instances combinations of adjustments are the best choice. It is difficult to generalize since it depends on the specific product produced as to which approach is best.

An important case where an integrated system is essential for control is the extrusion of blown film. In addition to the elements that we have discussed, the blown film introduces orientation effects, air introduction for bubble-size control, variable freeze-line locations, effects of the physical properties of the set film on the line, and many other interacting parameters. A mathematical model of the system that permits analysis of the interactions is necessary so that a control strategy can be generated. In Ref. 4 a modeling procedure and the way of converting the model to a scheme for controlling the process is discussed.

Figure 10-3 from the reference shows the two systems and the parameters affecting the product as well as some of the interaction that takes place between the two subsystems. Figure 10-4 shows several of the feedback loops used to control the process to make the film fall within the gage specifications with the desired degree of orientation and clarity. Figure 10-5 shows the control computer scheme or flow sheet to operate the process. The nature of the on-line data inputs and the required stored information are shown as well as the way they are processed to supply the adjustment of the machine and systems control.

The film lines represent one of the more advanced systems for controlling extruders where the use of a computer is invaluable in generating optimum product at maximum rates. If the system is taken one step further with the requirement that the product is a coextruded material, the control is additionally complicated by the need to coordinate the several machines producing the melt streams and the die conditions necessary for proper bonding between the layers of material. Here the model is more complex and the degree of control required to make the product is much higher. The temperature and pressure of the melt at the combining point must be controlled to permit proper bonding

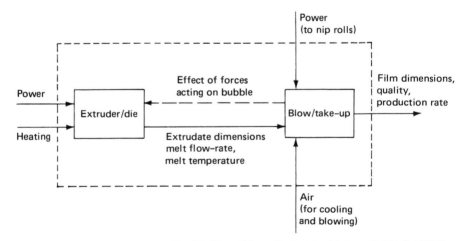

Reprinted from Polymer Engineering & Science **15,** *711, Oct. (1975).*
Fig. 10-3. Block diagram for blown-film extrusion system. The system consists of two subsystems: (a) extruder and die, (b) blow-up unit and take-up units.

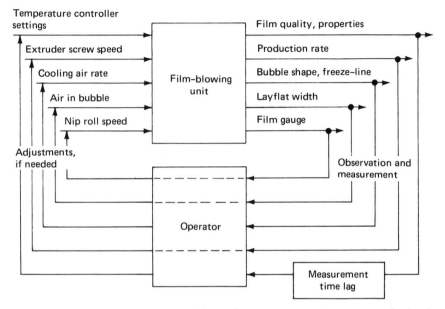

Reprinted from Polymer Engineering & Science 15, 712, Oct. (1975).
Fig. 10-4. Feedback loops and operational control points for blown-film extrusion system.

and a minimum of interfacial shear to obtain good bonds and smooth layers of material.

There are additional advantages to using computer systems for control of the extrusion process. One is that data can be recorded on the product and the process. Product data records can be used to provide a basis for a continuous quality monitoring system. Checks on stored information will indicate what machine conditions were at a time when the product went out of specification and can provide data for correcting the problem. In addition, the records can be used to indicate whether the product was out of specification at the time it was made or whether there was a shift in dimensions due to handling or postextrusion distortion. The data-logging activity should be designed to provide whatever data records are needed for troubleshooting and quality control. It can also be used for management and overall plant control by indicating downtime and causes.

System Hardware

Implementation of the control systems downstream as well as in the extruder requires sensors that can accurately monitor the process and produce signals that can be processed to operate the process controls.

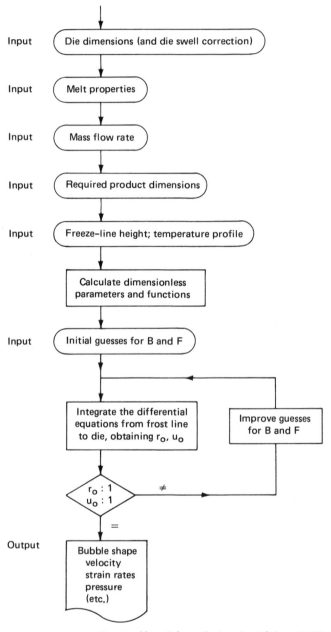

Reprinted from Polymer Engineering & Science **15**, 720, Oct. (1975).
Fig. 10-5. The computational scheme for control of the blown-film extrusion process. The flow chart shows the sequence of computer operation used in the process-control computer.

The system for the extruder uses thermocouples or resistance temperature devices to measure the temperatures and either a relay or SCR unit to modulate power or coolant to the machine. Pressure is read with a strain gage unit coupled to the melt with a fluid-filled tube. These sensors and controls are well known and widely used. Some of the measuring devices downstream for dimension and product quality are worth discussing to show how they operate and how their outputs are used.

One additional need in examining sensors is to determine if they can produce spurious outputs and, if so, under what conditions. If poor signals of dimensions are introduced, they can cause the system to destabilize. As can be seen from Table 10-1 gaging heads to measure dimensions range widely. Mechanical gages are among the simpler devices, but they have two major limitations. One is that the roller contact will smooth out dimensional variations smaller than the roller. The other drawback is the inertia of the device. With a rapidly moving line the inertia may be such that the device will lag the surface and give erroneous signals.

Air gages are a semimechanical gage type. Since they have much lower inertia, and generally lower time constants, they can read a dimension better than a roller contact device. They are usually operated against a reference signal by moving the gage to produce a constant air back pressure, and in this way can avoid errors due to nonlinear calibration. Figure 10-6 is a schematic of a rod line using both an air-gage sensor and a rolling contact sensor. The system illustrates the situation that frequently occurs where the size monitoring device must be located near the die in order to effectively control the process without excessive

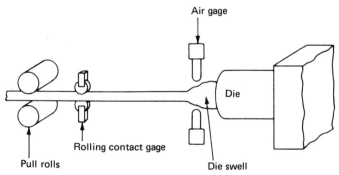

Fig. 10-6. Layout of sensors for measuring and controlling a rod extrusion line. An air gage sensor is used at the die for diameter control and a rolling contact gage is used, after the rod is hardened, to measure the diameter.

signal lag. The dimensions gaged near the die are not close to the actual sizes that the profile or section will be when it is pulled and cooled. This necessitates the use of another gaging device located near the end of the line to record the actual size of the section. The readings from the two gages can be used to determine the effect of variations in the cooling-line parameters on the size, and enables the operator or the control computer to correct for takeoff-equipment deviations.

Figure 10-7 shows how the output from the air gages can be used to make a closed control loop to the puller for size control. In this instance, there is a direct feedback with a signal processor to convert the dimension variation in the rod at the die to the appropriate voltage variations for the SCR motor control to adjust line speed to correct for size variations. A more sophisticated system would also use the input from the rolling gage and some data from a memory describing the fluctuations to be anticipated plus a software program that would correct the dimension. The computer control would replace the signal processor and the computer might also change some other line parameters such as the spacing between the die and the cooling units, temperature of the cooling agent (air or water), or some other element in the line to effect better control.

The air gage has the advantage of simplicity in construction and operation in the hot environment close to the die, but it has the disadvantage that the air released may affect the product by localized cooling.

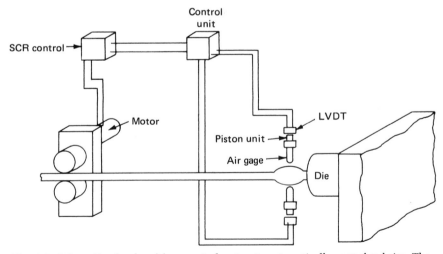

Fig. 10-7. Schematic of a closed-loop control system to automatically control rod size. The control sensor is an air gage driven to constant gap and the displacement read with an LVDT. The error signal is used to control the puller speed with an SCR drive.

Other probes such as the capacitive sensor shown in the sketch of Fig. 10-8 can be used. The probe is part of a circuit that is an RF oscillator that can be tuned or detuned by the proximity to the plastic surface. It is relatively rugged and it will not affect the material passing by. It also has a shorter response time and can be used to sense fluctuations that occur in shorter intervals along the extrudate, especially if the extrudate is going at high lineal speeds. In addition, the capacitive gage can be operated either as a direct but nonlinear gap dimension measuring device or as a constant-gap device with the displacement monitored by an LVDT (linear voltage differential transformer) and the device moved by a solenoid driver to the constant gap. The signals produced by either method can easily be used by the control or recording devices.

Another type of proximity sensor is shown in Fig. 10-9. This is a retroreflecting photodetector light unit. The light source used is generally a light-emitting diode (LED) or other small light source focused to a point on the moving extrudate. If the surface moves away, the photodetectors indicate the change and the linear actuator rebalances the unit with the resultant displacement indicated by the LVDT.

Another type of optical proximity unit is shown in Fig. 10-10. This is two intermingled bundles of fiber-optic fibers, with one-half going to a light source and the other half going to a photodetector. The backscattered light level is a function of the proximity to the surface with the nonlinear response curve shown. By measuring on one side of the minimum or the other, it is possible to determine the distance from the probe to the surface as a signal from the photodetector. This unit can also be operated in the constant distance mode with the linear displacer and LVDT output.

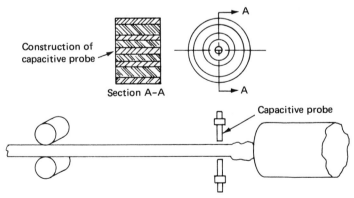

Fig. 10-8. Schematic of a capacitance proximity probe and the arrangement used in extrusion gaging.

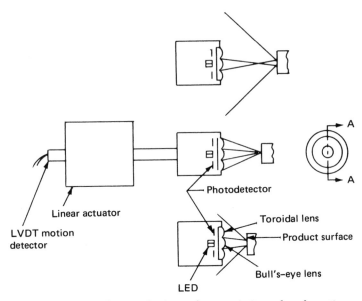

Fig. 10-9. Optical retroreflective surface proximity probe schematic.

Other surface location units can be used in the control systems for dimensions, but the ones recited are typical. The other dimension usually required is the thickness of an extrusion. This is true for tubing, sheet and film, profiles, wire covering, and a wide range of extruded products. The units used to sense these dimensions are of several types, with ultrasonics, nuclear gages, and magnetic gages among the ones most widely used.

Figure 10-11 illustrates the use of a magnetic Hall-effect gage. It is a gap-measuring device and, in order to be effective for thickness measurements, the air gaps need close control. An area of application for this unit is in making tubing with a mandrel sizing by the use of an internal sizing element. In this operation the tube is snug against the sizing mandrel, and the outside gap can be kept to a constant value with a suitable air bearing design. The calibration curve of Hall-effect voltage versus gap is linear over the range of use for a particular gaging operation, and the signals produced are readily handled by the control and recording equipment.

Capacitive units can be used as thickness as well as proximity measurement devices. Figure 10-12 shows a thickness gage setup indicated on a tubing wall which can be used equally well on sheet or film. A simplified method of incorporating the unit into a puller control system is

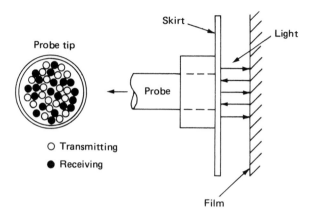

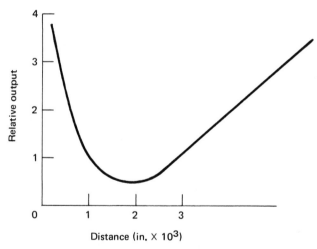

Fig. 10-10. Optical surface proximity gage using fiber optics. One-half of fibers bring light and the other half bring the reflected light to the photoelectric sensor. Note the calibration curve.

also shown. As with the other control systems this is the simplest feedback method and can be replaced by the use of microcomputer control systems.

Ultrasonic gaging is widely used in controlling pipe and tube wall thickness because it is one of the best one-side thickness sensing means. The use of an ultrasonic wall thickness unit is illustrated in Fig. 10-13.

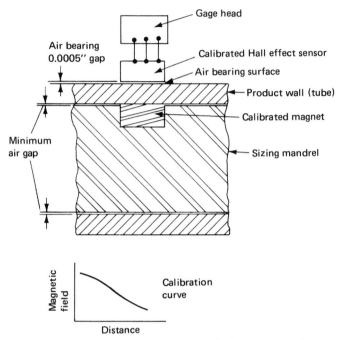

Fig. 10-11. Schematic arrangement for the use of a Hall-effect sensor and magnetic field to measure wall thickness. Internally sized tubing extrusion is the process shown. Note calibration curve.

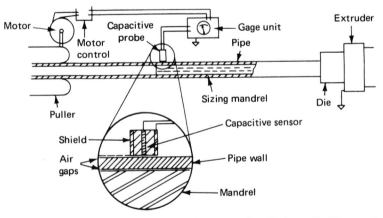

Fig. 10-12. Schematic of tubing extrusion line showing gaging of tube wall with capacitance gage.

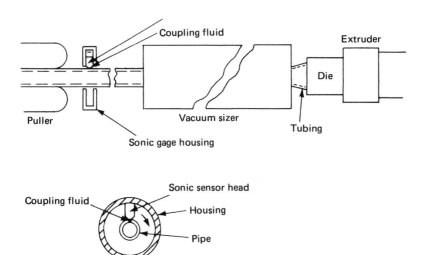

Fig. 10-13. Pipe extrusion line schematic showing pipe wall thickness gaging using reflected ultrasound. Gaging head rotates around pipe to measure wall on all sides.

There are several operating modes for the ultrasound, including a variable frequency signal method, which uses the beat frequency between the transmitted signal and the return signal as a measure of the time of transit through the wall and a pulse-echo system, which measures the transit time through the wall as a measure of the wall thickness. Both actually derive the thickness from the acoustic velocity of the material and the time of transit. The pulse-echo system can also indicate the presence of voids or imperfections in the wall from echoes produced from inclusions. Another datum that can be produced by a dual-frequency ultrasonic unit is a measure of the quality of the plastics material in the wall. This results from the differences in acoustic velocity for different material conditions at different frequencies.

The nuclear gages include x-ray devices, beta-ray devices, and other forms of high-energy-radiation monitoring units. The detectors are special to the specific radiation form, and include gas-type detectors and scintillation units as well as charge decay units. The thickness measurements are made by measuring the particle or ray absorption of the material measure, and require the use of calibration methods using known thicknesses of the material gaged to calibrate the detectors. The units can measure through the material or they can be designed as one-side measuring devices. The most common of the one-side measuring devices uses backscattered beta rays to measure the material. The primary source passes through the material and hits a metallic back plane or surface. The primary electrons knock electrons out of the metal surface which, again, pass through the material and are measured by a suitable detector on the same side of the wall as the source.

One of the unique characteristics of the nuclear gages is that they measure the capture cross-section of the material which is directly related to the mass per unit area of the web measured. This leads to some interesting problems and some unique results in its use for gaging. For example, the fact that plastics are compressible materials leads to a situation in making sheet where the sheet is denser than normal due to processing conditions. In this case the gage will show greater thickness than the actual thickness of the material.

The density sensitivity of the nuclear gages leads to the successful operation of the Autoflex system of sheet thickness control. The system is illustrated in Fig. 10-14, and the construction of the thermal bolt lip adjusters is shown in Fig. 10-15. The control unit is intended to adjust the thickness across the width of the die with the die bolts and in the machine direction by speed control of the web. Unfortunately, since it is generally not possible to put the sensor between the die lips and the roll stack, the sheet has its dimension set by compression between the rolls reducing the transverse dimension variations to a minimum. However, the squeezing action of the rolls on the thicker sections of the sheet

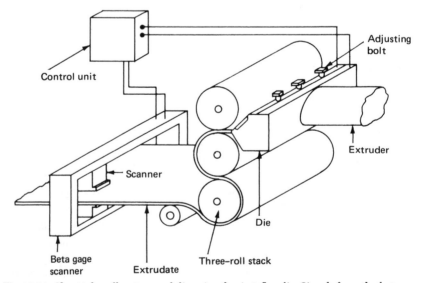

Fig. 10-14. Sheet take-off system and die using the Autoflex die. Signals from the beta gage scanner are used to control die lip opening to maintain constant dimension across the width of the sheet.

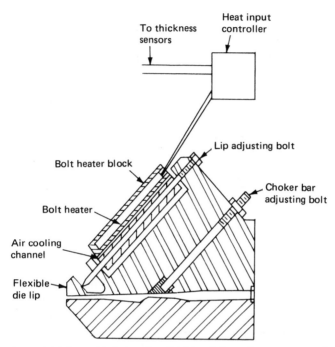

Fig. 10-15. Diagram of thermal bolt Autoflex die. Signals from the gage are used to vary the current flow in the bolt heaters which control bolt length. Variations in the bolt length will open or close die gap locally.

increases the density and, since the nuclear gage measures the area density of mass, it can still read the variations in sheet thickness coming from the die but in an attenuated manner. The system works so that the sensitivity of the gage is adequate even under these difficult operating conditions. If a computer is used in the system, the control can be improved by taking into account the effects of the roll stack on the final dimension readings.

The Autoflex system, which has as its main control the dimension of the die lips, is interesting as a working unit. The same approach can be used on other systems using other types of dies. Adjustable orifice tubing and wire-covering dies are made, and it is also possible to make profile dies with adjustable openings. Figure 10-16 illustrates one way in which such an adjustable die can be made. The usual operator unit to move the die elements in this type of unit is a cylinder connected to an electroservovalve with high-pressure hydraulic fluid connected to the cylinder.

The hardware described represents a selection from the wide variety of units available to operate a controlled production system. The computers used range from minicomputers, which can be used to control

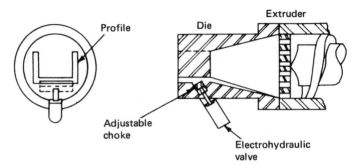

Fig. 10-16. Adjustable profile die using restrictor choke actuated by an electroservo-valve-controlled hydraulic cylinder. Shape control of the profile is possible with such units.

several lines, to small dedicated microcomputers which are used for a single system. Programming of the computers is essential to their specific use, and the cost of such programming is a significant element in the cost of a system. The advantage of the computer systems is the flexibility in control that they offer as compared to a simple feedback control.

The economic justification of the on-line control systems is based on the amount of usable product that can be made. Its main advantage lies in the fact that the reject level for product is lower with close control, and the tighter the quality standards, the easier it is to justify the use of elegant control systems. A secondary savings area is illustrated in Figs. 10-17, 10-18, and 10-19. By narrowing the swing in dimensions of the

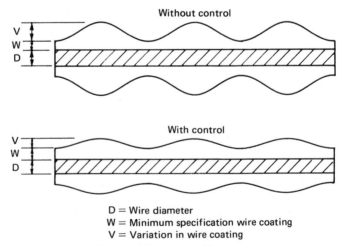

D = Wire diameter
W = Minimum specification wire coating
V = Variation in wire coating

Fig. 10-17. Sketch of the coating-layer variation in wire coating using automatic on-line control as compared with not using the control.

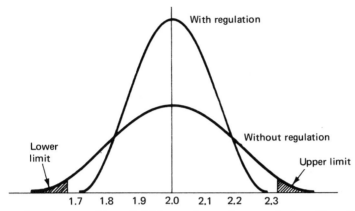

Fig. 10-18. Gaussian spread in wall thickness of extrusion-coated wire with and without regulation or automatic control.

wire coating it is possible to meet the minimum coating requirement by shifting the center of the dimension distribution to a lower value. In this way there are savings in both the material and the machine time—assuming a constant material throughput on the equipment.

Since the sources of savings are related to the reduction in the amount of resin processed to produce a given amount of product, it is apparent that cost justifications are based on the volume of product made. This is

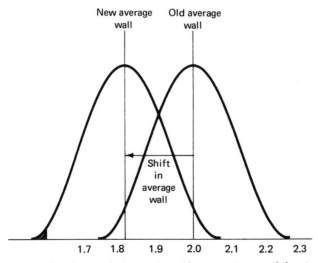

Fig. 10-19. Gaussian distribution of wire coating with automatic control showing the shift in the curve peak possible by the use of automatic wall thickness control. The shift represents the savings in material and machine time.

not true for the case where manual control of the process is inadequate to produce any usable product within specifications. In all other instances the justification is measured by the material and machine-time savings over a specific operating period.

A consequence of the savings base is that high-volume operations, such as the production of pipe, sheet, and wire, are those that justify the best installations. For example, a large sheet line will make 1500 lb/hr of product. The operating cost may be $150/hr and the material cost $0.35/lb. This represents a material and operating cost of $0.45/lb. If the system can reduce the amount of material in the sheet and the amount of reject to represent a 10% increase in average output per pound, the cost savings would be $150 \times 0.45 = \$67.50$/hr. For 7200 hr of operation per year this saving would be $468,000/year, which would justify a large investment in control systems. Typical pipe production systems can have 5% savings in material and time at the same scale of production which, again, justifies extensive control expenditures. In general it is true that virtually any system that is operated on a continuous around-the-clock basis can readily justify the necessary control equipment. The more difficult justification is with short-run product lines where there is a large start-up cost per run and where the cost of the control equipment to handle a large range of product variation is much more costly to build and to program.

Summary

The discussion has shown where and how on-line control with closed-loop control can be very effective in controlling the extrusion process. Strategies for control of the extruder, the downstream system, and the integrated system indicate the points of sensing the process parameters and the methods of control. Some system hardware is discussed to show how such control is implemented and how the data gathered by sensors are used in the control of the process. Examples of specific systems show how the control has actually been used in practice. The basic economic justification of the control systems as compared to manual control is demonstrated to be based on the amount of useful product that can be made per pound of polymer processed. The justification for high-throughput continuous operations is easily made, and, in general, the control systems are justified for any continuously run operation. The justification for custom operations with short runs of a wide variety of products is much more difficult and in many instances the use of sophisticated systems control, except for extruder output control, is difficult to justify on an improved productivity. Products requiring close control of dimensions can be justified even in this instance on the basis of being able to produce the product at all. Manual operation in this case will produce high enough reject levels to make the process unfeasible in the real sense.

The concepts are described with respect to specific examples, but are applicable to any extrusion system. Selection of the appropriate control units and the best control mode will require careful analysis of the product and system parameters. The use of mathematical modeling of the system can be used to determine the key parameters and the control points for a new system.

References

1. J. Parnaby, A. K. Kochhar, and B. Wood, "Development of Computer Control Strategy for Plastics Extruders," *Polymer Engineering & Science* **15**(8), Aug. (1975).
2. S. Levy, "Gaging and Control of Extrusion Processes," *Plastics Machinery & Equipment* **8**(6), 31, June (1979).
3. S. Levy, "On-Line Gaging of Extrusion Dimensions," *Plastics Machinery & Equipment* **7**(1), Jan. (1978).
4. C. J. S. Petrie, "Mathematical Modelling and Systems Approach in Plastics Processing: The Blown Film Process," *Polymer Engineering & Science* **15**(10), Oct. (1975).
5. A. D. Schiller, "Omnyson Correlations with Process Conditions," Haake, Inc., Technical Bulletins TB 2002, 2003, 2004, and 2005.

Chapter 11
Plant Design and Operations

The extrusion operations that have been discussed to make the various products utilize a group of machines cooperating to manufacture the desired extrusion. These machines require a plant that has suitable space requirements for installation of the equipment as well as for the services needed to operate the machines. In addition, the plant must have provision for raw-material handling, finished-product handling, and the other functions needed to manufacture the product. The operation procedures used must be known in order to operate the plant and to have appropriate personnel available. Appropriate maintenance facilities and personnel to keep the plant in order are a requirement of the plant operations; also required are appropriate health and safety procedures and equipment to avoid personnel injury and equipment damage. This chapter will discuss these aspects of the extrusion process in order to provide useful insight into designing and operating extrusion plants.

Plant Design

Plant design is concerned with the selection of appropriate equipment for production, layout of an efficient production line, design of a suitable building or structure to house the operation, and layout of the lines to bring the necessary services to the production line. In general, the equipment selection is done by the engineering and manufacturing engineering personnel who do the product development. Equipment is selected for the process to make the desired product of pipe, sheet, film, etc., which will consist of the extruder, dies, and downstream equipment such as pullers, cooling units, and cut-off units. The plant designer's function begins with the equipment list of the line components along with the requirements for space to house them and the necessary services to operate them, such as power, water, air, etc.

The first design problem is the layout of the equipment. A number of different layouts are used and these are based on the operation. Many operations such as pipe, sheet, and profile extrusion are in-line operations with the successive units arranged in sequence in line with the extruder barrel. The placement of the downstream equipment is determined by the length of the units involved and the space required between the successive units needed for stretching, cooling, and other in-air operations. In most cases the line width is generally not over 5 ft (1.5 m) and the overall line length will be determined by the cooling requirements and the equipment lengths. Lines can be as short as 25 ft

(7.5 m) (including the extruders) for a small sheet line to as long as 200 ft (61 m) or more for a high-speed tubing or profile line. The major line difference is the distance taken by the cooling equipment.

Some tubing lines are crosshead operations and all covering operations are crosshead operated. This produces a much different equipment layout with some space complications. Figure 11-1 is a diagram of an in-line operation and Fig. 11-2 is a layout of a crosshead covering operation such as a wire- or cable-covering line. The line is set up at right angles to the extruder barrel and the arrangement is an unequal leg T formation. It is clear from the illustration that a plant with a number of covering operations will represent a problem in efficient space utilization. The output from the several lines will interfere with each other especially if long cooling units are necessary. Figure 11-3 is a layout for several in-line operations that is widely used. Figure 11-4 is a layout of a multiline covering operation. It is evident from a comparison of the two that much more floor area is necessary for the covering operations to avoid interference between the operations.

One way to overcome the problem is to use 45° crosshead lines or use offset die lines. The effect of using these alternatives is shown in Fig. 11-5. The 45° offset lines make the plant more compact by turning the machines and enabling the payoff and downstream units to be placed parallel to each other. The offset die arrangement brings the operation back to an in-line operation with the payoff units behind the extruder. The offset lines will generally be somewhat wider than a simple in-line operation, but they are far more efficient in the use of space than the crosshead lines.

There are two other directions for the extrudate to come away from

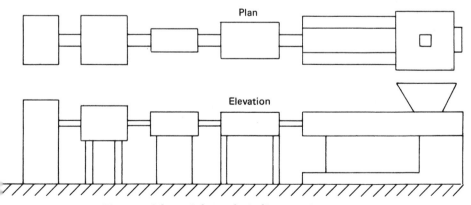

Fig. 11-1. Schematic layout for in-line extrusion systems.

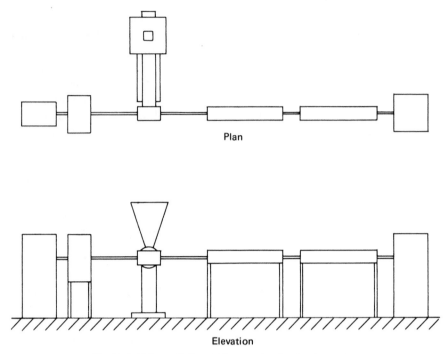

Plan

Elevation

Fig. 11-2. Schematic layout for crosshead extrusion system.

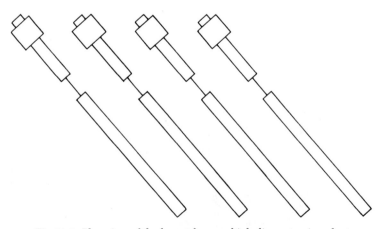

Fig. 11-3. Plan view of the layout for a multiple-line extrusion plant.

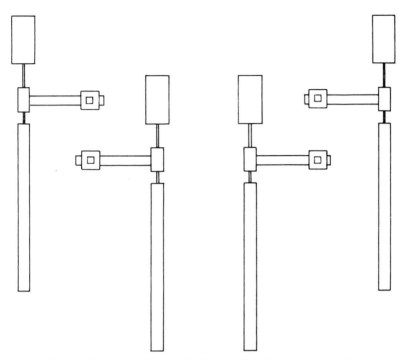

Fig. 11-4. Plan view of a multiple-line crosshead extrusion operation.

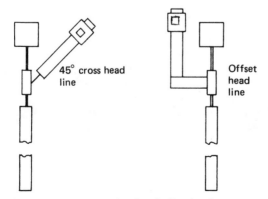

Fig. 11-5. Plan view of a 45° crosshead and offset-head extrusion operation.

the machine, up and down, and these directions are used in making
blown film, cast film, coated stock, and monofilaments. In the typical
blown-film operation the material is extruded upward into the cooling
and take-up tower. This arrangement is shown schematically in Fig. 11-6
and in the picture in Fig. 11-7. Depending on the scale of operation, the
tower can be as high as 20–40 ft (6–12 m) while the floor space
requirements are relatively small. The obvious requirement for high
headroom space is apparent. Some forms of blown-film systems extrude
down, and the extruder is on an upper deck in the plant with the tower
below the machine.

Another down-directed operation is the one used for film casting and
stock coating. This arrangement is shown in the photograph in Fig. 11-8.
The operation illustrated is on one floor because the die is fed overhead,
but in many cases the output is fed through the floor to a collecting
arrangement placed below the floor level of the extruder. The space

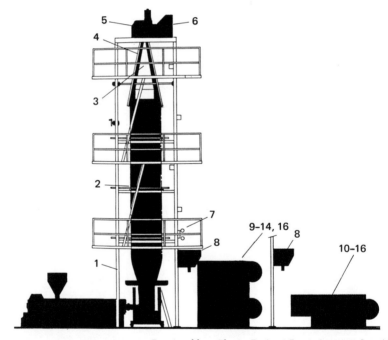

Reprinted from Plastics Design & Processing, *p. 43, June (1974).*

Fig. 11-6. View of multistory blown-film line. 1, Tower; 2, bubble guide; 3, gussetting
assembly; 4, collapsing assembly; 5, upper nip assembly; 6, treater assembly; 7, web guide;
8, dancer assembly; 9, winder; 10, slitter stations; 11, lower nip assembly; 12, lay-on roll
assembly; 13, dancer roll; 14, automatic indexing; 15, single turret winders; 16, winding
mandrels.

Courtesy *Egan Co.*

Fig. 11-7. Diagram showing operations at different levels in a blown-film extrusion operation.

Fig. 11-8. Slot cast film operation arrangement.

requirements for this operation are less than for sheet since the collection and cooling units do not require much floor space.

Monofilament lines are usually equipped with down-directed tooling, but in this case the lines are long in the machine direction because of the requirement for orienting the filament under controlled conditions. A line layout with some typical dimensions is shown in Fig. 11-9. After the downward immersion in the cooling tank, the filament is handled horizontally to stretch-orient the product. Similar arrangements are used to make oriented strip and ribbon.

The special requirements of the extrusion operation must be considered for the building or structure to house the different operations. Long buildings are needed for in-line operations, especially high-output lines, to provide room for extensive cooling equipment. Depending on the amount of powder used, there is the need for storage areas for the material—a warehouse for bags and boxes of material or silos for bulk material. Operations such as blown film require either high bay area buildings or multilevel buildings to contain the towers used to handle the film. This is true whether the operation is a conventional up-direction system or a down-direction system. In addition, the building must have facilities for unloading raw materials, for finished product storage, and

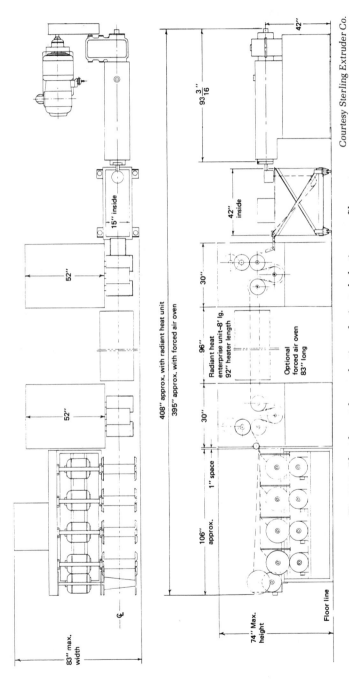

Fig. 11-9. Plant layout for production of oriented plastics monofilament.

Courtesy Sterling Extruder Co.

loading docks for shipping the finished material as well as space for the ancillary operations such as inspection, maintenance, tool maintenance and repair, and other functions such as clerical and supervisory activity.

Services

There are requirements for various services for the operation of the plant. The following is a list of those commonly required:

1. Power—generally heavy three-phase 440 V plus 110 and 220 V single phase
2. Water for cooling
3. Sewerage lines for wastewater disposal
4. Refrigeration systems for cooling the process water
5. Compressed air for operations and for operating auxiliaries
6. Exhaust systems
7. Buss duct systems for process data lines

Extruders are generally electrically heated and driven by large electric motors. Typical power requirements for an extruder will range from 121 to 134 hp (90 to 100 kW) for a 2½-in. (6.35-cm) machine to 670–1340 hp (500–1000 kW) for 6- and 8-in. (15.2- and 20.3-cm) machines used for large sheet and pipe lines. Present practice is to use 480-V three-phase in-plant distribution systems to the machines to reduce the cost of the wiring. This voltage is stepped down for the auxiliary and control equipment. Some of the voltage change is done at centrally located transformers and some is done at the machine.

The power distribution to meet current safety regulations is either distributed by overhead buss ducts or by under-the-floor ducts. The buss duct overhead is usually more convenient if frequent changes are needed to attach new or different pieces of equipment. With the wide-spread use of SCR drives one additional electrical arrangement used is an isolation transformer for each large extruder motor drive. The sensitivity of the SCR power control systems to line transients makes this arrangement necessary to prevent noise from other equipment on the mains triggering a shutdown of the SCR drive.

Water is required for cooling the extruder and is usually needed in other cooling equipment on the line. In most plants the cost of the water precludes dumping it into the sewer after one pass through the cooling system. Some water is contaminated by the processing and may have to be dumped for safety reasons or because of potential corrosion problems resulting from the breakdown products of the polymer. PVC operations with water cooling frequently must have a portion of the water dumped to avoid the buildup of hydrochloric acid. For this reason, sewers must be equipped with suitable water-treatment equipment to neutralize the contaminants in the effluent. The recycling will minimize the sewer needs and this will have a substantial effect on plant operating costs.

The recycled cooling water is cooled in several different ways. The one selected will depend on local conditions and local ordinances for water handling. One method is to use refrigeration systems to cool the water. This is a power-intensive technique which, again, will increase plant operating costs. The justification for its use will be based on sewer and water costs and dumping limitations under local codes. It can be cheaper than the costs of the services plus the required sewage treatment. Another cooling method involves the use of cooling towers where the water is cooled by heat transfer to the air and by evaporation. Obviously this method is weather dependent and is best used in cooler climates or where the relative humidity is normally low. This permits effective cooling by either heat transfer or evaporation. This method is to be avoided in hot, wet climates. Another method uses the groundwater as a cooling agent. Large finned heat exchangers are buried below the water table and the process water is passed through them to dump the heat into the ground. A water layer is desirable to move the heat away from the heat-exchanger location. Selection of the specific process water cooling will depend on demand requirements and cost.

Compressed air has three prime purposes in the extrusion plant. It is frequently used in tubing operations to control the size of the tubing and for air cooling of the extrudate. It is also used to operate air cylinders or air motors which are a part of the extrusion system. The third use is for signaling or measurement instrumentation. Normally, the plant air is produced in a central compressor area with one or more large units. The air should be cleared of oil and water for any of the applications. A minimum of particulate filtration is required for air motor or cylinder applications. For applications in sizing and cooling, 50-μm filters should be used to prevent clogging of the units and contamination of the product. If instrument air is used, the filtration requirements are much more stringent. Cooling and sizing air should have 20–25-μm filters and the instrument air 10-μm filters or better as designated by the equipment manufacturer. Since it is costly to do the filtering because of pressure loss and the cost of filter replacement, it is a good idea to filter locally only that air needed for the component requiring it.

Exhaust systems are used to vent the gases evolved from processing some plastics materials. The polymers using these systems most frequently are PVC, polystyrene, acetal resins, acrylic resins, polycarbonates, and other materials whose breakdown products are noxious or toxic. The exhaust systems should be designed to resist any corrosive action from the breakdown products. In addition, the effluent gases may have to be treated before discharge into the air to meet local air-pollution standards. For example, the effluent from a PVC operation will require the use of a neutralizing tower to remove HCl from the gases before the effluent can be discharged. Other installations may call for carbon absorbants or burners to remove hydrocarbon and other products from the exhaust.

Frequently, larger plants use central process monitoring and control. In these plants it has become standard practice to use special buss ducts for bringing the signals from the machine instruments and the control information from the controller to the machine. The advantage of the ducts is that they shield the instrument wiring from noise generated in the plant and eliminate spurious information.

Plant layouts must locate efficiently the source of the services needed as well as the lines needed to carry the services to the operation. In almost all cases, efficiencies can be achieved by reducing the lengths of the service lines to minimize losses due to voltage and pressure drops.

Material Handling

Raw material and finished product movement and storage are important considerations in designing an efficient plant for extrusion. The flow of material through the plant can expedite the production, reduce personnel requirements, and also reduce the hazards of operations. Selection of the handling systems is important with respect to an operation since the tradeoff between capital requirements and labor requirements can affect the costs of production.

Raw-material storage and handling depends on the size of the plant and the variety of plastics processed. Raw material is delivered by truck or railcar, and it can be packaged in bags, drums, or large boxes. In larger operations the material is delivered in hopper trucks or hopper cars. The bags, boxes, or drums are usually on skids or pallets and are handled into a warehouse by using forklift trucks. This is a labor-intensive operation, and when large amounts of a specific material are used, bulk deliveries are preferred. The bulk materials are unloaded with air conveyor systems into storage silos. This operation is capital intensive, requiring the construction of the storage silos and the installation of the air-conveyor systems to unload the hopper units.

Delivery of the material from storage to the machines will be different for the two systems. Usually the packaged plastic is moved to the machine by forklift trucks and dumped into the hopper by hand or with small hopper loaders. This, again, is a labor-intensive operation and requires space at the machine head end for some local storage of material, usually one or more skids. The silo-storage operation generally has air-conveyor distribution to the machines. This is a capital-intensive arrangement, but it does have the advantages of close control of material going into the operation and the lack of secondary storage requirements at the machine. The central silo and distribution systems are easily justified in sheet and film operations with one or two grades of material and for pipe and profile using one or two materials. It is very difficult to use with custom operations that need less than 40,000–50,000 lb/run (18,160–22,700 kg/run).

If the main change between runs is the color, the silo scheme is still desirable, with the addition of color-addition feeders on the machine. The color changes required at the machine are simple to make, and the main polymer storage system will handle the bulk uncolored material.

Finished product handling and storage is more diverse, and the handling and storage is peculiar to the specific product line. Pipe and heavy tubing are usually cut into lengths and bundled or boxed. The handling to the storage area is either by conveyor or forklift truck. Heavy sheet is cut and stacked on pallets and, again, depending on the plant size, is moved to storage either on conveyors or by lift trucks. Thin sheet, film, light tubing, and wire are examples of products rolled up at the end of the extrusion line. These reels are moved with lift trucks or carried by conveyors to storage. In some instances, for very large reels, they are hand rolled to the storage area. Palletizing of the reels is common, and this is usually done in the warehouse area.

The typical finished-goods storage area is a rack set up on which the pallets of finished goods are stacked. Some plants load full-trailer-length sections of the pipe directly into the trailers to minimize handling of the product. Thus the warehousing is done in the trailers.

Shipping from extrusion plants is usually by truck. Rail is used on heavy production items such as pipe where it is available. Appropriate dock facilities are necessary in the plant for trailers and railcars if used. The use of conveyor equipment is common for loading heavy pallets.

Plant Layout

It is not feasible to cover all of the processes with regard to an appropriate plant layout because of the diversity of product and the lines used to produce the products. Figures 11-10 and 11-11 are layouts for a blown-film operation and a pipe plant, respectively, and they show the space assignments for the production areas and the other auxiliary work

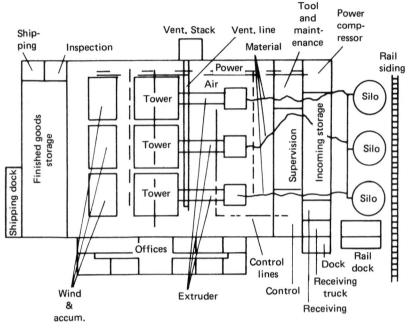

Fig. 11-10. Plant layout for the manufacture of blown-film showing location of equipment and auxiliary functions and operations.

areas. This type of plant layout should be prepared for a contemplated production plant in order to determine the space requirements and building type as well as to locate effectively the services and service lines for the process. By estimating the plant output the dock, storage, shipping, and handling areas can be determined. The equipment placement can be done to ensure both a flow of materials through the plant and that the necessary inspection and control functions are located in the best part of the operation.

Operations

The operation of an extrusion plant involves a number of tasks of which the actual machine-and-line operation is the most important. The additional functions performed are inspection and quality control, tool set up, tool maintenance and repair, material control, and general supervision of the operation. In addition, there are engineering activities involved in new product tryout, new material tryout, and equipment improvements which must be checked on the full-scale system.

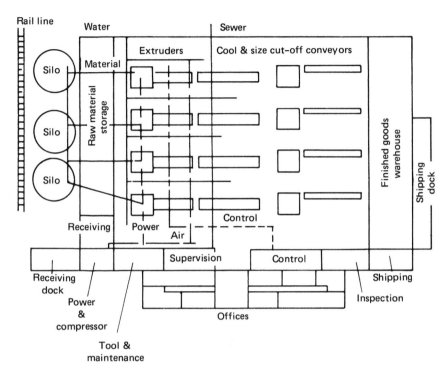

Fig. 11-11. Plant layout for the manufacture of plastics pipe. Equipment layout and auxiliary functions are shown as well as the material handling and storage systems and shipping docks.

Inspection and Quality Control

This is one of the important elements in the operations. If there is not a well-organized and well-run inspection and quality-control system in the plant the scrap generation of off grade and size products can be so costly that it can shut down the operation. The inspections can be done automatically with on-line gaging equipment, or they can be done manually by the operator. In either case it should be checked by inspectors who periodically measure and test specimens of the production to ensure that the line gaging is correct. Batch control, another function performed by the quality-control operation, enables the inspectors to trace specific production lots to raw material supplies as well as to the time the material was run, which is necessary for the investigation of field complaints on off-specification material. The problem can be traced either to materials that are out of specification or to improper operating conditions in the equipment. The quality-control and inspection department is also responsible for incoming material control. Even if there are no facilities for testing the raw materials, the retention of samples from each batch of material received is essential. In the event of a problem, the samples can be sent to a laboratory for tests, but they must first be available.

Scrap control is another quality-control concern. This is a factor that might arise from off-grade product or from process control problems that occur with totally useless machine output. The source for discarded product information is generally the quality and inspection operation. This is relayed to management for correction.

Maintenance and Set Up

Generally these are not regarded as associated tasks but in the context of an extrusion operation they are usually performed by the same people. Some setting up of lines and die changes are done by the operating personnel and the line supervision. However, when a line is taken down and replaced with a new setup when the product is changed, the maintenance and set-up crew does the work. In addition, the crew is responsible for the periodic maintenance work on the equipment and the die repair and cleanup.

Preventive maintenance on extrusion equipment is necessary for smooth plant operation. In any large-volume operation the equipment is operated on three shifts and, frequently, over the weekends. A breakdown can lead to substantial downtime in the plant. This is especially a problem when the item that failed is one for which spares are not stocked. A motor drive failure can result in shutdowns of a week or more. As a consequence, planned maintenance and a program of stocking spares difficult to order on short notice is important for good plant operation.

Die repair is one aspect of the maintenance that is difficult to manage. Frequently, to avoid loss of production, dies are run past the point where they should be pulled for major rework. Figure 11-12 shows a sheet die removed from operation after severe damage necessitated a shutdown of the lines. Without attention to the requirements for tool maintenance, the only time the dies are pulled is when the truck ratio becomes unity, i.e., the rate material is returned from the customer is equal to the rate at which it is shipped. Since major tooling, such as sheet dies and blown-film dies are expensive units, it takes understanding of the maintenance needs to induce management to get enough extra tooling to permit the removal of tools from service periodically for overhaul to bring them back to proper operating condition. This reduces the scrap level and frequently improves the productivity to the point that the extra capital cost is readily justified.

Fig. 11-12. A well-abused sheet die shown after disassembly, showing broken bolts and damaged die surfaces caused by usage. The disassembly and repair will cost a substantial fraction of the cost of a new die.

A unit desirable for cleaning dies is shown in Fig. 11-13. This is a fluidized-bed cleaner which removes plastics materials from the die by the action of sand or ceramic miniballs on the metal in a heated fluidized bed. This method reduces the nicks and scratches likely to occur when the cleaning is done by hand. If the tool cleanup is done by hand, wood

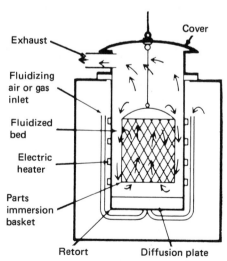

Courtesy Procedine Co.
Fig. 11-13. Diagram of a fluidized bed used to clean dies and other extruder parts. Heated sand or ceramic microballs are used in the device (Procedyne).

or very soft metal tools should be used. Hard metal, even carefully handled, can result in tool damage.

While the maintenance and repair of the electrical and electronic equipment is not strictly a tool-maintenance operation, it is usually under the activities of the maintenance group. This involves drive maintenance on the extruders and other machines as well as maintenance on instruments and controls. A good instrument mechanic is essential to smooth plant operation, especially if the plant has extensive feedback control on the operation. It is now necessary for the instrument repair personnel to have some training in maintenance on microcomputer systems installed as part of the control systems. The level of repair may only extend to localizing the malfunctions to a particular circuit board which can then be replaced and sent to the supplier for repair. Here, too, a spares program should be used to have available those replacement parts critical to the maintenance of production. The spares program should be matched to the available skill level. If subsystem replacement is necessary, the spares should be geared to this. If the capability is servicing the subsystems, then the spares can be essential components to wire into the boards.

Supervision

Line supervision in an extrusion plant involves scheduling of a sufficient number of operators to cover the operation, plus those needed to provide supplies of material and to remove finished product. The number and skill levels needed depend on the operation. If the operation is highly automated with central material supply by air conveyor and automatic product collection as well as on-line controls for feedback maintenance of dimensions and other product qualities, the number of people required is small—one operator could probably handle two or three film lines or two or three pipe lines. This operator's main activity would be to monitor the operation of the automatic equipment. In a number of operations the operator would also be the inspector since little other activity would be required. The skill level of such an operator is high, and the need is for well-trained individuals who understand the process.

Less automatic lines require someone to adjust machines for output. They also require material handlers to load the machine hoppers and to remove finished product from the factory floor to the warehouse. The inspectors are usually other individuals who have sampling test responsibility, while the machine operator also checks the product to be sure that the equipment remains in adjustment.

On manual lines scheduling for regular breaks is also necessary. In many cases an automated line will operate so well that the operator can leave for reasonable periods of time without causing any problem. This is not the case with most manual lines because they need close continuous attention to avoid scrap and other operating problems. Operations with intermediate degrees of automatic production will be scheduled to supply personnel that can do the necessary work to maintain production.

One important aspect of the supervision is scrap management. Any material that does not become product in the first pass through the line must be carefully handled to avoid contamination. In some cases the material can be reground and returned to the process as a limited percentage of the total throughput. In other cases the material must have the additives increased for stabilization before the resin can be rerun even on a limited basis. In many cases the product requirements are set so that none of the scrap can be rerun into the same product, as, for example, products covered by FDA regulations. In these cases the material must be disposed of to a scrap dealer or run for some utility product with low material specification requirements. The proper handling and storage of scrap is necessary to a smooth running operation.

In addition to the handling of the scrap, it is a supervisory responsibility to minimize the generation of off-grade and scrap product. This is done by determining the sources of the scrap and by modifying the operating procedures and control systems to reduce the amount produced. A typical source of scrap generation is the material produced on the start up of a line. Until the process is brought under control, all of the product is discarded as scrap. The start up scrap is of several types. The initial output is usually improperly melted and in some cases may be contaminated. This material should be segregated from the next run material which will have erratic dimensions but will be of the same quality as the good product. After the process is running it may take some time to reach stability in output, and a part of the production will be off size and need to be rerun. The latter two groups of material should be carefully handled with respect to avoiding contamination since in many cases this material can be recycled by just grinding.

In order to cut down the scrap, the start up sequence should be monitored and improved as much as possible. This can be done by careful note taking on each startup to determine which sequences bring the process under control most quickly. In the case of a computer or other closed-loop control system, the data on the start up can be stored in the controller for a programmed start. It must be determined which process parameters have the most effect on dimensions, and they should be carefully monitored to determine the operating sequence which most rapidly brings the dimensions to the required values. Again, with on-line gaging and control this information can be programmed into the control so that the line will correct dimensions rapidly and maintain them in production.

Training operators in scrap management is one of the most important supervisory responsibilities. The percentage of scrap is one of the best ways to monitor productivity. In fact, the scrap level is one of the main determining factors in decisions regarding machine production rates. Incrementing up the output at the expense of increasing the amount of unusable material is usually counterproductive.

General approaches to scheduling for a continuous around-the-clock operation are important in high-volume extrusion operations. Because of the startup costs and losses, continuous operation is necessary to extrusion plants. The second and third shifts are frequently problem times with lower productivity. Having some of the best troubleshooting personnel on these shifts is important to continuous production. This is typical for any continuous operation and is important for extrusion due to its complex character.

The organization of the operating personnel is usually to have operators responsible for one or more machines in terms of controlling production, auxiliary floor personnel (usually experienced operators to fill in for breaks and to assist when the operators need assistance), and floor foremen who cover the floor activity and are generally the set-up men for turning on the lines and changing the product lines. Inspectors are usually attached to the quality-control function and report directly to management. Maintenance and repair personnel are usually under the plant engineering activity and are on call for equipment replacement and repair. They are usually involved in major equipment shifts for line set ups and installation of new lines and equipment. The control activity in plants with centralized control systems is usually directly under the plant or operations manager, who reports the general information on productivity and equipment utilization. Usually the manager is also responsible for scheduling on the equipment and for the determination of work load and lead times for production.

Two engineering functions relate directly to the extrusion operation. These are tool development and the tool test required on production equipment. In a fixed product operation such as film or pipe production this is done at a relatively low level. In the operation of a profile or other specialty product operation the tool engineering and tool testing are a major part of the operation. It is the key factor in determining the work load. It is important to monitor the tool testing carefully. Typically, the hourly cost of operating a line on tool test is about twice that of normal operations. Typical hourly rates are $75-$125. With this type of cost structure plus the cost of rapid tool modifications, the cost of new tool development can run to tens of thousands of dollars with a difficult product. In many cases the product potential does not justify such expenditures and the programs ought to be reviewed and terminated if the tool development time is excessive.

The other general responsibility in the engineering activity is manufacturing engineering. This is the type of effort dedicated to improving the performance and output of the production lines. Additions of better controls and line units to increase output are generally the major activity. Assistance in tool design to help tool designers make tools fit into the operations is another activity of the manufacturing engineers. They also design and build the on-line cutting and shaping units used, such as punches, formers, and similar equipment.

In summary, the operations of an extrusion plant require a mix of personnel that depends on the nature of the operation. The procedures for operation should be clearly developed and followed to permit smooth functioning of the production lines. Support personnel in control, maintenance, inspection, and engineering are needed to complete the overall plant complement to handle all of the manufacturing activity needed.

Safety and Health

An increasing concern for plant safety and health conditions has been accentuated by the activities of two government agencies—OSHA (Occupational Safety and Health Agency) and EPA (Environmental Protection Agency). In addition to the usual safety awareness that is common to all good plant operations, it is necessary to comply with the regulations of these agencies. In a number of instances the compliance is counterproductive. The regulations do not, in fact, add to job safety and may increase the danger in some operations. It is important for plant safety personnel to recognize this and to actively counter the adherence to regulations if they are not viable. If they comply and there is an accident, they are still responsible even if they did what was required by the regulations.

In several areas the extrusion operation does present some hazards. Frequently, the pressure in the extruders is high and under some malfunction conditions can be dangerously high, leading to equipment rupture. Loss of the head end from an extruder does occur, and it is possible that this can lead to a large piece of steel moving like a projectile across the room. It is imperative that no one stand in front of an extruder in the start up phase because it is the most likely time for this accident to occur. In all cases it is wise to lay out the line so that if there is a head-loss failure, it will be deflected by downstream equipment to prevent serious injury to personnel.

Another hazard area in extrusion is the electrical shock hazard related to the connections made to the dies. In most installations 220-V AC is used to heat the die components, and these heaters are exposed on the outside of the machine. Even with good connections to the heaters that are protected from accidental contact, there is a problem, usually as a result of a grounding of the heater internally or as a result of damage to the heater terminals and leads. Water is usually present around an extruder, and wet floors are very common. The 220-V electrical current and the wet floors present a potentially lethal hazard. Every effort should be made to prevent an accident by good housekeeping and careful attention to repair on the die heaters.

Many of the units on the line can present potential hazards which must be protected against. Cut-off units with flying knives must be guarded to prevent access to the knife locations while the machines are operating. Cut-off saws need guards which will keep hands away from the saws during operation of the cutoff cycle. If there are on-line fabrication operations involving presses and similar units, the appropriate guards and interlocks are required.

The pulloff units represent another safety consideration. Three-roll stacks for sheet lines are now required to have rapid open arrangements that actuate easily from a safety bar to prevent severe injury to someone threading the unit. Pullers are now required by OSHA regulation to have the approach and exit sections blocked so the operator cannot get his hands in between the pinch rollers or belts. Larger pullers are required to have a rapid open feature with a safety trip bar. In many cases these additions to the puller can create more risk than they prevent, but the accepted standards require their use unless they are specifically exempted for good cause.

The other consideration in factory working conditions is the air itself. Some of the plastics produce fumes of various types that can be noxious or toxic. In most operations the level of effluent is low, but there is a requirement that the fumes of materials such as styrene, PVC vapors, acetal resin vapors, polycarbonate resin vapors, and many other materials vapors be exhausted through hoods over the die section of the extruder. In the case of large-scale use of materials such as PVC, there is a requirement for a health-monitoring program due to the potential liver cancer problem.

Exhausting the air from the factory is also complicated by the requirements of EPA. Many of the effluent gases must be absorbed from the exhaust stack before the air can be exhausted. This adds a significant cost to the exhaust system requirements. In each case the specific code requirements should be checked and appropriate scrubber equipment installed. For low effluent processes it is sometimes possible to obtain waivers on the exhaust system requirements, but usually only in small operations.

Water used to cool the product directly is another new problem in terms of regulation. If there is any problem with regard to the cooling water having contaminants that may cause health problems, the access to the cooling water by operating personnel must be restricted by such means as the use of covers on the cooling tanks. Acetal resin is a material which has been indicated as a potential problem. In addition, under current EPA regulations, it is necessary to monitor the water that is discarded to the sewer and to eliminate any potentially damaging materials by proper treatment of the wastewater. Vinyl chloride is one material strictly controlled.

There are many other health and safety regulations that apply to plastics processing operations. It would be advisable to have someone in the organization designated to follow the regulations that apply to the plant operations and to initiate appropriate compliance with the regulations. This could be the plant safety director who would also be responsible for normal safety measures and review of the safety equipment and procedures required for plant operation.

Operating Costs

While this is a section mainly involved with plant design and operation, there is a general need to know the costs that determine the hourly rate required for a machine. This information is useful to guide the operations manager in handling high-cost portions of the operations, and it is essential to management in cost work on new products. The following list of the elements that make up the costs for a 2½-in. (6.35-cm) extruder will be a useful guide in determining machine hourly cost. Table 11-1 summarizes the costs for the machine and the elements. While not directly comparable for other operations and machine sizes, it will be present in any cost accounting for this purpose.

Hourly Cost Calculations for 2½-in. (6.35-cm) Extruder

Space costs: Typical space requirements for a 2½-in. (6.35-cm) machine with in-process material storage is 1000 ft² (92.9 m²). Present rental costs are in the range of $2.50–$3.50/ft²/yr. Using the average of

Table 11-1. Hourly Costs for Machine

Space costs	$ 0.40
Power costs	2.50
Sewer and water	0.5625
Cooling	0.76
Maintenance	0.54
	$ 4.76
Machine amortization	$ 1.42
Total costs except labor	$ 6.18
Typical labor rates with fringe benefits is $6.50/hr, and a properly run operation should use one man plus 10% for relief time to make the cost per hour	$ 7.15
Total Hourly Rate	$13.33
Supervision costs about 20% since one supervisor can handle five machines with a supervisor cost of $15.00 which will come to	$ 3.00
Machine Time Cost	$16.33/hr

$3.00, the total annual cost is $3000 with an hourly cost based on 7500 hr of operation of $0.40/hr.

Power costs: A 2½-in. (6.35-cm) extruder is normally equipped with a 50-hp (37.3 kW) drive and around 30 kW in heater load, which makes a total extruder load of 80 kW. The downstream units usually add about 5 kW to the total load. Since the heaters only operate intermittently when the extruder is delivering near full output, the assumption is that the unit will draw the 50 kW for the motor and 25% of the heater load of 7.5 kW plus the downstream load of 5 kW, which comes to a total of 62.5 kW. At current prices power averages $0.04/kWhr, to give an hourly cost of $2.50.

Sewer and water costs are for makeup cooling water: A typical rate is $2.50/1000 gal for water and $2.00/1000 gal for sewage. Typical usage would be 1000 gal/shift to make the cost $4.50/8 or $0.5625/hr.

Cooling costs: As an example, cooling costs represent the costs of chilling the cooling water in the cooling tanks. It is expected that the total power input to the extruder is removed as heat in the cooling tanks. The extruder power is 57.5 kW which represents the heat input. Using a cooling unit with an efficiency factor of 3, the power requirements of the cooler would be about 19 kW, which represents an hourly cost of $0.76.

Maintenance: The best way to set the cost of maintenance is to add the number of hours for routine maintenance, such as oil level, greasing, etc., to an estimated time for repair based on experience. The time may be 2 hr plus 40 hr/yr to make a total of 154 hr. A typical maintenance hourly rate would be $20. Spares for such things as instruments, motor control parts, etc., should be kept on hand. An estimated cost for repair items such as $1000/yr is typical. The total maintenance cost is then about $4000/yr which would be $0.54/hr.

Summary

This section has been concerned with the plant design and operation which shows the interaction of the operational needs in the plant with the layout and design of the facility. The range of operations based on the extrusion process are diverse, so that no one plant arrangement or set of operating procedures is applicable to all of the operations. The examples given in the chapter illustrate some of the typical arrangements and operations to guide in the arrangements for a new facility and for the improvement of existing plants. Safety and health considerations must be taken into account in both design and plant operation. With the basic material presented it will be possible to intelligently undertake the necessary operation and plant design.

Chapter Twelve
Extrusion Products and Processes

The extrusion process for plastics is capable of producing a wide variety of products. Some of these are relatively simple in shape and structure like sheet materials and some have complex structures and unique characteristics. Variations in materials and in processing are used to create interesting products. Some of the products and processes were described specifically as examples in describing the technology of the extrusion process. In this section we will examine the different materials and options in the process, and indicate how they can be used to make unique items.

Orientation

One interesting characteristic of polymer-based materials is that they normally exist as relatively disordered structures, but they can be oriented by processing to align the polymers in the process direction. Other variations in the process produce biaxial orientation that aligns the molecules in a plane. In either case, the effect is to produce materials with substantially different properties from the unoriented polymer.

One significant improvement is that the tensile strength and modulus of elasticity are increased. Normally, the orientation introduced by stretching the extrudate under closely controlled conditions will increase the tensile strength by 3–10 times with stiffness increases that range from 2–5 times. A process actively under development—ultraorientation—can produce tensile strength increases of 20–50 times and increases in the modulus of elasticity of 50–100 times.

The production of blown film results in orientation in the film. Control of the blow-up ratio and freeze line will produce material with different levels of orientation. High draw-down ratios in extrusion of rods and profile shapes will also introduce a substantial degree of orientation. The arrangements are shown in Figs. 12-1 and 12-2 from Ref. 1.

Figure 12-3 shows a biaxially oriented sheet made by using a tentering conveyor. The material coming from the sheet die is grabbed along both edges with traveling clamps and carried into an oven where the sheet is conditioned to the orientation temperature. At this point the sheet is stretched laterally by the clamps as they move along the track and longitudinally by the forward speed of the clamps increase over the rate at which the sheet is extruded. The resulting biaxial orientation, usually 3:1 to 4:1, produces sheet with several times the tensile strength of unoriented sheet and approximately three times the stiffness. One

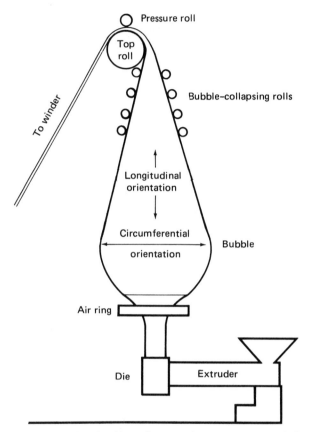

Reprinted from Plastics Machinery & Equipment *8 (11), 30, Nov. (1979).* ©
Fig. 12-1. Orientation effects produced in blown-film extrusion.

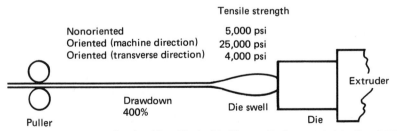

Reprinted from Plastics Machinery & Equipment *8 (11), 30, Nov. (1979).* ©
Fig. 12-2. Orientation effects produced by extrusion drawdown.

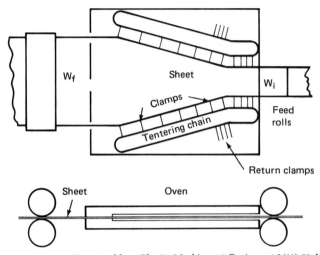

Reprinted from Plastics Machinery & Equipment **8** *(11), 30, Nov. (1979).* ©
Fig. 12-3. A tentering frame and oven arrangement used to produce biaxially oriented film or sheet.

widely used sheet material processed in this manner is made from polyethylene terephthallate polymer (Mylar). Many other polymers can be similarly biaxially oriented and now some polyolefin materials are made this way.

Figure 12-4 shows the line used to make oriented strip and monofilament. The material is extruded and quenched in a cooling tank. Then it is stretched controllably by the use of differentially driven Godet rolls after

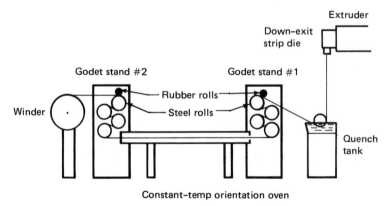

Reprinted from Plastics Machinery & Equipment **8** *(11), 30, Nov. (1979).* ©
Fig. 12-4. Monofilament and strapping orientation line using Godet stands for stretching the extrudate in a constant temperature orientation oven.

the material is brought to the optimum temperature. High-tenacity monofilament with tensile strengths of 50,000 to 150,000 psi (344.5 to 1033.5 MPa) can be made from polymers such as polyolefins, nylons, thermoplastic polyesters, and others. The packaging strapping materials of nylon and polypropylene are made in the same equipment with slight variations. A unique version of the oriented strip is the plastic ribbon which is oriented foam strip. The strength of the material is high, and the foam cells give the product a lustrous silky appearance.

Orientation can be produced by rolling as well as by stretching. Figure 12-5 shows an in-line rolling operation performed on plastics extrusions. The severe reductions in section at each pass through the roll nips make highly oriented materials. Oriented nylon strip is made in this manner and the product can have tensile strengths of 50,000 psi (344.5 MPa) and higher with substantially increased stiffness characteristics. In Fig. 12-6 the process of rolling is applied to a simple profile where the enhancement of physical properties makes a unique product.

It was previously indicated that the so-called ultraorientation of the polymers produces much larger increases in physical properties of strength and stiffness. The ultraoriented material rivals aluminum and, in some cases, steel in stiffness and strength. One way these materials are made is shown in Fig. 12-7. The condition produced in the material is shown in Fig. 12-8. All of the polymer molecules are aligned to give maximum stiffness and strength. Another characteristic of ultraoriented material is that it will give a dead bend similar to that of annealed metal.

In addition to the method shown in Fig. 12-7 a number of other techniques are used to make ultraoriented extrusions. References 2, 3, and 4 describe methods that appear to have promise for producing these unique products. They will have a wide range of applications for

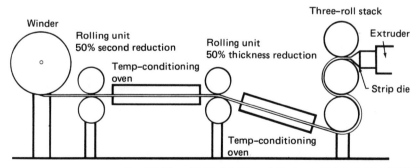

Reprinted from Plastics Machinery & Equipment *8* (11), 30, Nov. (1979). ©
Fig. 12-5. Diagram of processing line used to orient strip or sheet by rolling. A thickness-reduction of 50% at each set of rolls is typical.

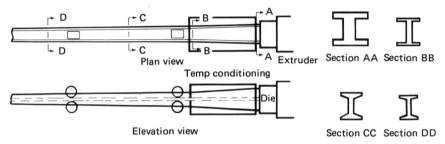

Reprinted from Plastics Machinery & Equipment 8 (11), 30, Nov. (1979). ©
Fig. 12-6. Roll orientation line used on simple I-beam section.

high-strength members as well as for reinforcement in other matrix materials.

There are other effects produced by the orientation process on the properties of the materials. For one thing, the oriented materials are denser and less permeable to both liquids and gases. This makes the oriented products useful as barrier materials in equipment and packaging applications. Another result of the orientation is that the materials will shrink in the oriented direction if heated to a temperature near the original orientation temperature. This effect is used in shrinkage banding materials, in shrink-wrap packaging film, and in heat-shrinkable tubing.

The highly oriented materials also have interesting optical properties. When the molecules are aligned, the material becomes birefringent (double refracting) and can be used to influence polarized light passing through the material. In addition, some oriented films can be converted to absorption-type polarizers. This can be done with polyvinyl alcohol by heating the film to remove some of the hydroxy groups from the polymer structure. In other cases the oriented material is dyed with a dichroic dyestuff that will produce the preferential absorption of the light vibrating in one plane.

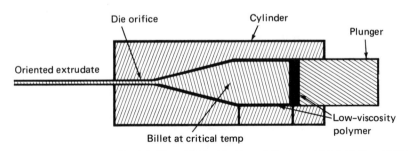

Reprinted from Plastics Machinery & Equipment 8 (11), 32, Nov. (1979). ©
Fig. 12-7. Ram extruder unit used to produce ultraoriented rod. The system shown uses a low-melt-viscosity polymer as a combined pressure transfer medium and lubricant.

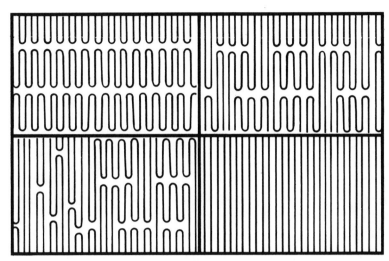

Reprinted from Plastics Machinery & Equipment *8* (11), 32, Nov. (1979). ©
Fig. 12-8. Schematic representation of the molecular pattern that exists in ultraoriented
polymer materials.

These are some examples of products that can be made by extrusion
and orientation. The unique character of the materials and process
enable the products to be produced for special effects. There are other
characteristics of oriented materials, such as changes in electrostatic
properties, and others used in even more esoteric applications such as
electrets.

Mechanically Operated Die Heads

Another group of products is made by the use of special dies which have
movable sections. One of the simplest of these is a pipe die with either
the mandrel or bushing, or both, turning. It has the effect of circumferentially orienting the polymer in the wall of the pipe and increasing the
burst strength of the pipe. This is mentioned in Ref. 5 and described in
detail in Ref. 6. Close control of the operating conditions can impart
significant improvements in properties. Figure 12-9 shows a die that is
used.

Another product making use of a rotating die mandrel is the spiral
coaxial cable material shown in Fig. 12-10. The die used to make the
product is shown in Fig. 12-11. The mandrel has the web section cut into
its section and is connected to the main flow through the die. By rotating
the mandrel the web is swept around between the inner core containing
the wire and the outer tube of coaxial covering. The technique has also
been used to make twin-walled tubing for heat exchangers.

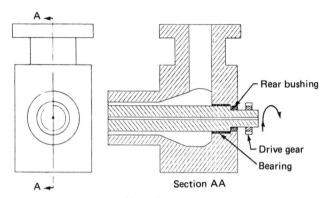

Reprinted from Plastics Machinery & Equipment **8** *(5)*, *44, May (1979).* ©
Fig. 12-9. Rotating mandrel tubing die which is used to make circumferentially oriented plastics tubing.

Some other interesting tubular products can be made by operating the pin or mandrel in a die axially. In Fig. 12-12 the mandrel is equipped with a series of stop-offs to the flow. Tubing with a perforated wall can be made by oscillating the pin in and out in the direction of flow. Figure 12-13 shows a die structure that can produce tubing with a variable wall thickness. This type of head is used in programmable parison making for extrusion blow molding, and it can be used to make variable wall tubing for a number of applications. One system for making corrugated tubing uses the variable wall tubing to improve the wall thickness control in materials sent through the corrugator.

Netting is another product made with operating section dies. The die has either one or two movable sections rotating or oscillating past each other. The openings through the die alternately permit and obstruct flow

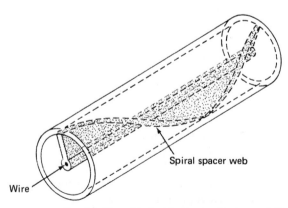

Reprinted from Plastics Machinery & Equipment **8** *(5)*, *44, May (1979).* ©
Fig. 12-10. Sketch of spiral separator web construction coaxial cable insulator structure.

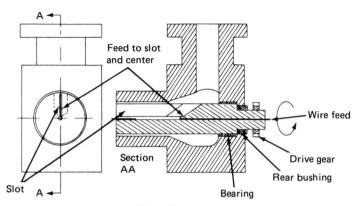

Reprinted from Plastics Machinery & Equipment **8** (5), 45, *May* (1979). ©
Fig. 12-11. Diagram of the construction of a rotating mandrel die to produce the spiral web
coaxial insulation shown in Fig. 12-10.

partially to make the open-work structure of the netting. Figure 12-14
shows the basic construction of the round and flat netting dies. Figure
12-15 shows the sequence of positions of the die lips in the flat netting die
that produces the overlap sections to join the netting and the single
strand portions that are the connecting sections of the netting. Figure
12-16 shows some of the detail in the construction of a round netting die.

Figure 12-17 shows the rotating intersection system used to make
another variety of perforated wall tubing. The material is essentially the
same as the netting, but instead of having small strands and large
openings, the strands are wide and the openings are small. Tubing of this
type is used in drip-and-soaker irrigation systems. By careful die design

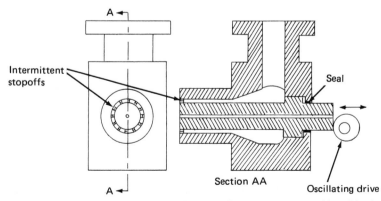

Reprinted from Plastics Machinery & Equipment **8** (5), 44, *May* (1979). ©
Fig. 12-12. Diagram of oscillating mandrel die which is used to make perforated wall
tubing.

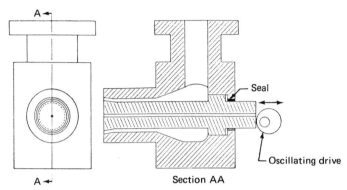

Reprinted from Plastics Machinery & Equipment *8 (5), 44, May (1979).* ©
Fig. 12-13. Diagram of a reciprocating mandrel tubing die which is used to make tubing with variable wall thickness.

the openings can be made very small so that the outflow from the tubing can be closely controlled.

Another interesting product is made with a rotating mandrel film die. Two different materials are introduced into the die, usually polystyrene and polyethylene, through alternate openings in the rotating bushings, and the resulting film is produced with a large number of alternating

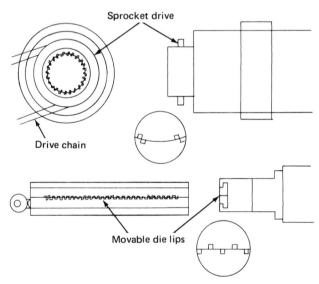

Reprinted from Plastics Machinery & Equipment *8 (5), 45, May (1979).* ©
Fig. 12-14. Diagrams of two types of dies with moving die lips that are used to make netting.

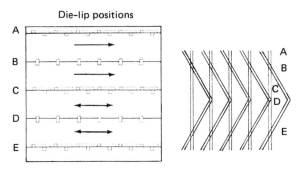

Die-lip positions

Reprinted from Plastics Machinery & Equipment **8** *(5), 45, May (1979).* ©
Fig. 12-15. Diagram of die actions in a flat netting die that produce the opening in the netting.

layers. The resulting film is iridescent with interesting optical properties including a deep pearl-like character. In addition, the alternating layers of the polymers introduce unique properties in the film because of the large number of resin interfaces. The character of the product is described in Refs. 7 and 8. The major application of the multiple-layer film is in packaging for both barrier pack purposes and decorative effects.

Twin-skin sandwich panels can be made by another type of active die. Figure 12-18 from Ref. 9 illustrates a cell-spaced inner core in a sandwich-construction die made of discrete cells produced by alternate application of vacuum and air pressure through the core-forming sections. The connecting webs are alternately deflected to one side and then to the other to connect the walls and produce a cellular core.

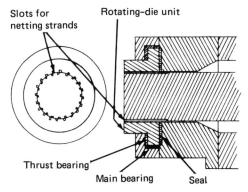

Slots for netting strands   Rotating-die unit

Thrust bearing   Main bearing   Seal

Reprinted from Plastics Machinery & Equipment **8** *(5), 45, May (1979).* ©
Fig. 12-16. Detail of the rotating bushing construction in a round netting die.

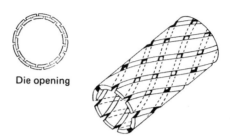

Die opening

Reprinted from Plastics Machinery & Equipment **8** (5), 45, May (1979). ©
Fig. 12-17. Detail of die opening in die to make perforated wall tubing by die and bushing rotation.

These are examples of the products that can result from the use of mechanically operated dies. Other dies of this type have been used to apply spiral reinforcement wires to tubing and other applications. These dies require careful design in order to operate properly in the high-pressure, elevated-temperature environment found in the extrusion process, and the seal design is critical. The tooling is relatively expensive and requires careful maintenance. The additional product range that it affords justifies the difficulty involved in these special dies and processes.

Products Based on Special Materials

There are a wide variety of products where the ability to extrude special materials into shapes is a key factor in the product utility. One of the simple applications is the use of glass or other fiber-reinforced materials

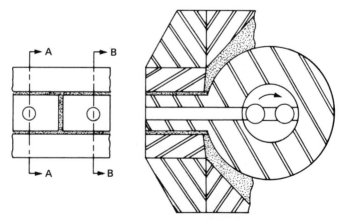

Reprinted from Plastics Machinery & Equipment **6** (12), 24, Dec. (1977). ©
Fig. 12-18. Cross section of oscillating valve die used to make honeycomb-type panel.

for structural members. The reinforcement can either be chopped fiber or continuous filament. One of the effects of the extrusion is to align the fibers in the extrusion direction making products with high unidirectional strength. With the continuous fibers the process is a continuous covering operation on the filaments yielding materials with 60-80% of the physical properties of the reinforcement material. Fishing-rod stock is just one example of how the product is used. Guy wires for large antennae are another application.

An interesting application for the process with a special material is in electrical conductivity applications. One widely used material for this purpose is carbon-black-filled polyethylene, another being polyvinyl chloride compound filled with carbon black. The recent availability of conductive-metal-flake materials has extended the range of electrical conductivity available at reasonable loading levels and broadened the scope of applications, which include among others static elimination applications such as floor tile for high-explosion-sensitive areas, conductive plastics profiles for handling integrated circuits for production, and static leakage sections in medical tubing for use with inflammable anesthetics.

There are several applications for the extruded conductive materials in electrical applications. They are used as lead wire for high-voltage, low-current applications such as ignition wiring. The materials are extruded as covering layers to act as shielding materials in other electrical cables. Frequently, they are used as coverings to bleed accidentally accumulated charges from antenna leads. They can be used wherever the application can tolerate resistivities from 10^3 to 10^6 times that of the metal conductors.

Magnetic materials can also be extruded and made into magnetic strips. These are used in door closer strips and in a wide variety of other uses. Usually the magnetic materials used are polyvinyl chloride compositions with high loadings of ferrites because they have generally good magnetic properties. The loading by weight is in the range of 75-90%. In order to make the magnets permanent the material is magnetized as it comes from the die. Figure 12-19 illustrates schematically how this is done. The material is extruded onto a drum which has a series of magnets arranged around the periphery with alternating north and south magnetic poles. The magnets are linked by a keeper ring so that the surface of the drum displays alternate north and south poles to the magnetic strip. The drum size is large enough so that the contact time will permit orienting the magnetic material in the strip to line up with the fields. The material is cooled as the drum rotates. The result is that the strip has impressed upon it a series of magnets with alternating north and south poles opposite to the configuration on the drum. The cooling sets

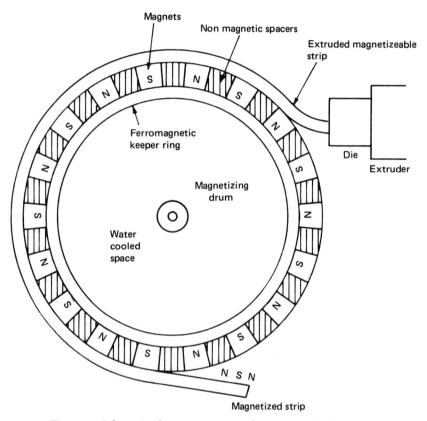

Fig. 12-19. Schematic of arrangement to make magnetized plastics strip.

these into the material. In some instances the magnets run parallel to the strip and the drum is constructed with parallel magnets.

By using specific polymers such as polyvinylidene fluoride plastic material and high loading of barium titanate and similar materials, it is possible to extrude plastics with very high dielectric constants. These materials have applications in special capacitors and in cables and lines for handling high currents without producing transients. There are other applications in electronics equipment when high electron storage characteristics are needed. The extrusion of these materials requires special techniques in the extrusion process since the rheology of the materials makes them very stiff even at the melt temperatures. Since the fillers are also abrasive, the tools and screws require extensive maintenance to repair the wear effects produced by them.

Filled materials are also the basis of other interesting extruded products. For many years all pencils were made from graphite writing material embedded in a wooden support. Now pencils are made by extrusion with all the sections made in one coextrusion operation. The writing material is a high-graphite-content plastic material which breaks down under pencil pressure to make the markings. The major part of the pencil body is made from a medium-density foam which has most of the characteristics of wood with respect to stiffness and sharpenability. The outer layer is a solid plastic material which gives a high gloss surface that is easy to handle and print on. The preferred basic material for foam and skin is polystyrene. The resin for the writing core is usually not identified, but is probably also styrene based.

Other writing materials besides graphite can be cored into this type of pencil structure, and colored pencils of all types are commonplace. In addition, the newer porous-tip pens use extruded porous materials as the tips for the pens. The extruded materials are rendered porous by a variety of proprietary techniques.

Some materials are important as coatings or coextrusions. This is especially important in medical applications where the surface character of certain materials such as the polyurethanes are necessary for making the surfaces nonclotting to body fluids. Coextruded tubing with a special polyurethane inner layer is important in kidney dialysis equipment. Similar tubing is used in the heart-lung machines to avoid clotting effects. Polycarbonate layers are also used in the oxygenating units in similar equipment. The materials that can be used for applications in medicine are highly selective. The structures range from tubing to filaments to special shapes, but the main performance characteristics of the devices are based on the special polymer characteristics ranging from surface chemistry to oxygen and other gas permeability characteristics.

Tubing and other shapes are used in other applications where permeability characteristics are important. For example, some of the gas separation membranes that can separate out oxygen are made from special silicone formulations extruded into high-surface tubing. Some liquid separation techniques are also based on similar materials.

Optical fibers are another product which is a materials-based application. The plastics optical fibers are generally made from acrylic resins. The major requirement in the successful production of an optical fiber is the layering of materials of carefully controlled refractive index on the outside of the main fiber section so that the light is conducted through the fiber in a coherent manner. Since the range of compatible acrylic resins with different refractive indices is wide, suitable layer materials can be found to make the fiber. The process involves coextrusion of a base acrylic with several graded index materials in controlled thickness layers around the fiber. Strict dimensional control is required because the dimensions usually have to be controlled within tenths of a micron. The material is usually extruded several times the final size and stretched to make the fiber the necessary small dimension of about 25–100 μm. Careful control over the purity of the materials is also critical to minimize the light loss by absorption in the materials. Plastics optical fibers have been made with properties closely equivalent to the quartz and glass fibers and at a much lower cost.

These are several examples of applications for extruded plastics in which the specific material characteristics are the key to the product utility. There are many others. The broad range of properties which polymers and polymer compositions can have lead to many interesting uses. The extrusion processing can enhance some of the properties and can convert the materials to useful shapes for many applications. The coextrusion processes can combine these materials to further extend the range of possibilities as, for example, extruding a magnetic tape with an adhesive layer so that it can be readily attached to a surface. The ingenuity of the plastics extrusion product designer will undoubtedly make many other applications possible.

Combination and Composite Products

The last cited example is only a small illustration of what is possible in extrusion processing in terms of combination products. In the coextrusion process for film and sheet, inner layers of adhesive are used as a combining technique for dissimilar materials. The adhesive layers can also be made external to the sheet of film to make a self-adhering material. The technique can be extended to profiles to make them adherable. In addition, it has been possible in several instances to also provide the separator layer to cover the adhesive layer with a resin readily separable from the adhesive. This suggests a whole series of potential applications where the slip or protective layer can be coextruded with the product.

The introduction of reinforcing webs into plastics strips, sheets, and profiles is another technique that can result in a variety of useful products. Reference 10 summarizes a technique and products made by this type of extrusion lamination process. The web is passed through a wide slotted crosshead die to combine the material in the extrusion machine with the substrate. Embedded metal mesh, coated board stocks, coated expanded metal, and coated cloth matrixes are a few of the possible combinations with applications ranging from ventilator structures to conveyor belting and formable metal embedment structures. One of the large-scale applications was for a fiberglass-based measuring tape which had a vinyl coating for stabilization of the glass tape and accepted printing for the indicia.

Several interesting composite structures are described in Ref. 9. Figure 12-20 shows one way of making a corrugated core sandwich panel material by the use of three extruders. The core can be the same material as the skins or it can be one that has good adhesion to the skins. The combining action is shown in a vacuum sizer, but roll methods have also been employed to combine the webs. These extruded sandwich products compare favorably with the longitudinally ribbed panels in Fig. 12-21. The combination units can be made with thinner material to make lower cost units.

Figure 12-22 demonstrates another combining technique to make a paper honeycomb core sandwich panel. Two thin-sheet streams are extruded on either surface of the expanded honeycomb as it passes through a crosshead unit. The composite material is useful as a structural member, and the material is frequently made with transparent or translucent cover layers to act as a light-entry-controlling glazing material in roof and high sidewalls in buildings. Another commonly used core is foam. Figure 12-23 demonstrates one way in which such panels can be made by a combining extrusion process. Essentially, the technique is a form of coextrusion with solid skins and foam interiors.

An application of the embedment technology in conjunction with the

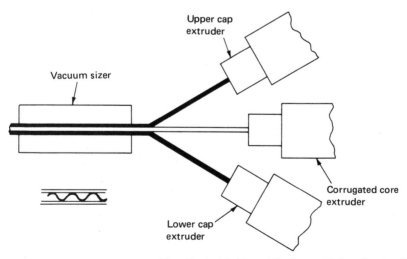

Reprinted from Plastics Machinery & Equipment **6** *(12), 22,* Dec. *(1977).* ©
Fig. 12-20. Three-extruder system to make corrugated core sandwich panel.

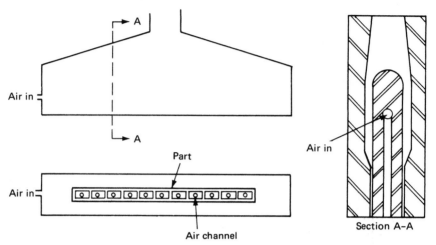

Reprinted from Plastics Machinery & Equipment 6 (12), 22, Dec. (1977). ©
Fig. 12-21. Extrusion die used to make hollow core panel.

manufacture of foam materials can supply a stiff structure that does not rely on the skin integrity. This is illustrated in Fig. 12-24 where a mesh layer is extruded into a foam section near each surface. The mesh acts the same way as reinforcing bars in concrete—by adding stiffness near the surfaces and making the section difficult to bend. Since the mesh is capable of extending under stress, the integrity does not depend on the skin as in conventional sandwich panels.

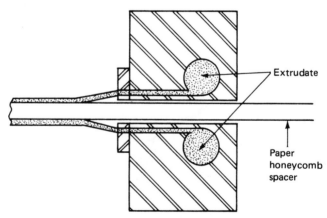

Reprinted from Plastics Machinery & Equipment 6 (12), 24, Dec. (1977). ©
Fig. 12-22. Schematic of die used to cover expanded honeycomb core with thermoplastics skin.

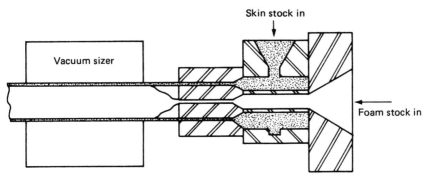

Reprinted from Plastics Machinery & Equipment **6** (12), 24, Dec. (1977). ©
Fig. 12-23. Schematic of die and sizer used to make solid-skin foam-core sandwich panels.

The combining technology can be extended to a wide variety of special applications. Figure 12-25 is an illustration of one interesting product. The skins of transparent plastic material have a series of thin metal strips extruded between them. The resulting product is a sunshade material which also has insulating properties created by the dead air spaces enclosed by the metal ribs and the cover layers. With the current emphasis on solar-energy control this represents a unique low-cost approach to heat-transfer control in glazing materials.

Downstream Operations

One of the applications for extruders is as a supply system for other types of plastics operations such as injection molding and extrusion blow molding. Because these methods rely only slightly on the general extrusion technology they would not be considered as primarily extrusion operations. However, there are a wide range of other hybrid processing

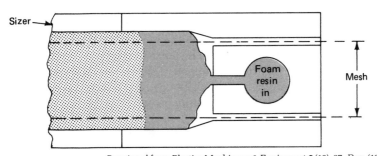

Reprinted from Plastics Machinery & Equipment **6** (12), 27, Dec. (1977). ©
Fig. 12-24. Schematic of die and sizer used to make mesh reinforced foam plastics panels.

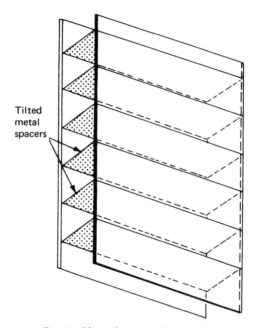

Tilted
metal
spacers

Reprinted from Plastics Machinery & Equipment **6** *(12), 27, Dec. (1977).* ©
Fig. 12-25. Die sketch and product made by using a crosshead extrusion system to make a
glazing panel with metal fin spacers.

methods which are predominantly extrusion processes where the on-line
operations significantly alter the nature of the product made. These
represent extensions of the extrusion processes.

One rather simple downstream operation done on highly oriented
polypropylene strip is an example of the variety of methods for downstream operations. The oriented strip has a tendency to split because of
the effect of weak interpolymer bonding between the oriented polymer
chains. The operation of fibrillation (a pounding effect) will convert
oriented strip to polypropylene staple yarn. The material is then woven
into sacking. Secondary operations on oriented products are one type of
on-line operation with interesting product potential.

Many of the other on-line operations are secondary forming steps
which reshape the constant cross-section extrusion into a material with
other shapes. Reference 11 describes a number of frequently used
operations. One of the simplest is the embossing of the extrudate for
decorative and other reasons including increasing the stiffness. Deep
embossing can also impart shapes to the part. The most complete
three-dimensional forming is done by vacuum- or pressure-forming
extruded sheet. Two techniques for accomplishing this are shown in Figs.

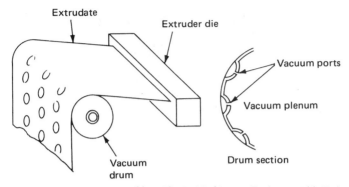

Reprinted from Plastics Machinery & Equipment *7* (8), 27, Aug. (1978). ©
Fig. 12-26. Sketch of the arrangement used to make deeply embossed sheet by the use of a vacuum drum.

12-26 and 12-27. The first employs a vacuum drum which pulls the sheet into shapes on the drum and is useful for small area parts. Drinking cups and small trays have been made using this method. Larger area parts are made with the chain-mold method in Fig. 12-27. The molds can be of fairly large size. One of the requirements is that the mold motion be intermittent or continuous to permit effective operation of the forming stage. One major reason for using the method is that reheating of the sheet for conventional thermoforming is avoided. While this saves the reheat energy, it also produces much better control on the heat in the sheet. Many materials that cannot be thermoformed in conventional machines can be on the in-line process. By removing just enough heat to bring the material to the forming temperature, materials such as propylene polymer can be formed. The difficult heating characteristics of this

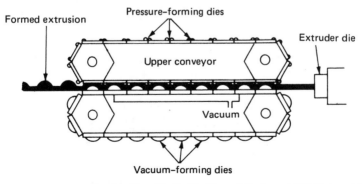

Reprinted from Plastics Machinery & Equipment *7* (8), 27, August (1978). ©
Fig. 12-27. In-line vacuum-pressure-former unit using a chain of dies mounted on a synchronized conveyor unit.

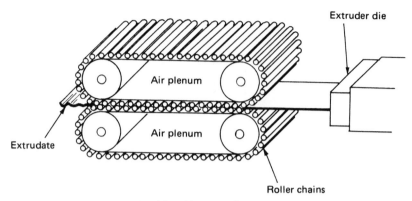

Reprinted from Plastics Machinery & Equipment 7 (8), 27, August (1978). ©
Fig. 12-28. In-line sheet corrugator unit.

type of material, combined with the narrow range between the melting and forming temperatures, makes normal thermoforming very difficult and chancy.

One-dimensional forming can also be done. Figure 12-28 shows how a wide simple profile strip can be reformed into a corrugated material by the use of an in-line corrugator and cooling device. Strips can be reconfigured to wave shapes of various types by means of a chain-forming device such as the one shown in Fig. 12-29. The shaping can be as complicated as required. One application is the formation of complex gasket patterns from flexible polyvinyl chloride materials for automotive seals and similar applications for building products.

The on-line forming operations can be much more complicated. Figure 12-30 shows a coiling operation. The most familiar uses for this product and technique are coil cords and coil tubing used in retractable handling of electrical and pneumatic equipment. Other applications

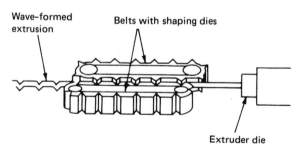

Reprinted from Plastics Machinery & Equipment 7 (8), 27, August (1978). ©
Fig. 12-29. In-line rod-shaper device.

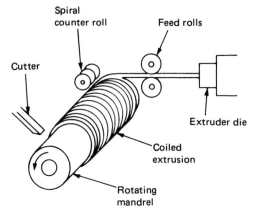

Reprinted from Plastics Machinery & Equipment *7 (8), 27, August (1978)*. ©
Fig. 12-30. Coil and cut-off unit for coiled extrusions.

have been made by this method with hot material in the form of special shape strip to manufacture large diameter or other large cross-section ducts and flexible hoses. The material is wound to form a helical structure when it is hot enough to easily weld together.

Cut-out sections on the extrudate are another way in which the process capabilities can be extended. The obvious methods are the punch presses and rotary die cutting. Other methods have been used to increase the possibilities. Figure 12-31 shows the use of a heated punch which can thermally cut or form the extrudate. Figure 12-32 shows how a laser can be used to similarly form the wall of an extrusion. The

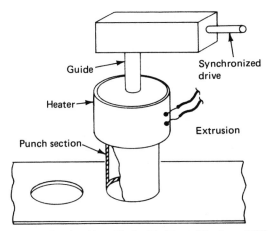

Reprinted from Plastics Machinery & Equipment *7 (8), 28, August (1978)*. ©
Fig. 12-31. Heated punch mounted on synchronized punching unit.

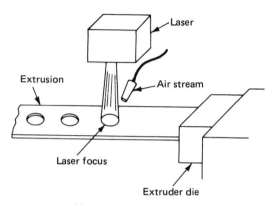

Reprinted from Plastics Machinery & Equipment **7** (8), 28, August (1978). ©
Fig. 12-32. Laser hole cutting unit for cutting accurate and clean holes in the walls of tubing and other extrusions.

justification for the use of a $125,000 laser is that the cutting can be highly selective and very clean. For example, a laser can cut through one wall of tubing or a multiwall profile. In addition, the shape is defined by nonwearing apertures in the optical system, which makes the cuts or shapes very consistent. The optical followers used eliminate the need for mechanical synchronization, and the cutting rates are high.

The reference describes many reforming operations on line exemplified by Figs. 12-33 and 12-34. In this operation a profile is twisted. Figure 12-33 shows the operation done with a rotating ring which is restricted to one size rod. Figure 12-34 shows the same operation performed by the use of a puller with crossed belts. This is a universal method that can accommodate any shape.

These are a few of the possibilities in making product variations by the use of on-line operations. These possibilities are seemingly inexhaustible. Combining extrusion and forming is one method for extending the diversity of products made by extrusion.

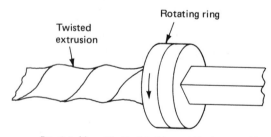

Reprinted from Plastics Machinery & Equipment **7** (8), 28, August (1978). ©
Fig. 12-33. Twisting bushing for making twisted rod.

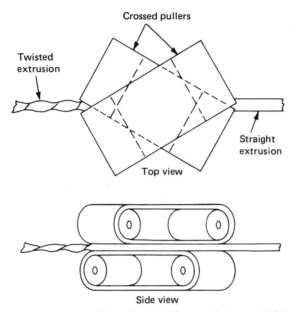

Reprinted from Plastics Machinery & Equipment *7 (8), 28, August (1978)*. ©
Fig. 12-34. A schematic showing how to make twisted rod by crossing belts on the puller.

Summary

This chapter gives an idea of the variety of product possibilities that have been explored using the extruder as the basic processing machine. Some of the products result from characteristics such as polymer orientation that can be imparted to materials by the process. Other products make use of special materials and combinations of materials to make new functions possible. Another group of products is the result of using special dies which produce regulated but nonuniform flows. Finally, the product capabilities are extended by downstream operations which modify both the shape and the characteristics of the extrudate. The methods discussed are appropriate to a wide variety of different extruder types with the proper machines used for the polymer pumping requirements.

The text has been intended as a basic guide to the plastics extrusion processes. A relatively simple machine in construction, the extruder has become a sophisticated device to produce a wide variety of products from plastics. An indication of its importance is the fact that over 50% of all polymer materials made are processed to final product in extrusion machines.

The basic flow characteristics of the polymer materials were reviewed to show how the data are used to design machines and tooling. Basic control methods were covered, as well as the more-sophisticated control systems now coming into widespread use in the industry. Specific extrusion lines such as those for film, sheet, tubing, and wire were analyzed to show the interplay between the extruder and the downstream equipment, and the approach to the design of a complete extrusion system discussed. Various special operations, such as coextrusion, were covered to show how they can be used to make special products combining the characteristics of several different resins.

The layout and design of extrusion plants were briefly described to show how the operation is assembled from the component line parts and how the operation is run. Manufacturing procedures and plant operations were discussed because of the impact they have on plant design and production and quality control.

The overview of the plastics extrusion technology should provide an insight into the process. No book can cover all of the detail needed. Indeed, some is not available because of the proprietary nature of many operations. References to relevant articles in the literature will provide additional sources of information for the plastics extrusion engineer, tool designer, plant engineer, manager, and others concerned with the extrusion manufacturing industry. It is felt that by providing this presentation the jobs involved in the industry will be made somewhat easier to understand and to accomplish.

References

1. S. Levy, "Improving the Performance of Extruded Products by Orientation," *Plastics Machinery & Equipment* **8**(11), Nov. (1979).
2. P. D. Griswold, R. S. Porter, C. R. Desper, and R. J. Farris, "Preparation of Ultra Drawn Polyethylene by a Novel Radial Compression Method," *Polymer Engineering & Science* **18**(6), May (1978).
3. A. E. Zachariades, P. D. Griswold, and R. S. Porter, "Ultra Drawing of Thermoplastics by Solid State Coextrusion Illustrated with High Density Polyethylene," *Polymer Engineering & Science* **19**(16), May (1979).
4. P. D. Hope and B. Parsons, "Manufacture of High Stiffness Solid Rods by the Hydrostatic Extrusion of Linear Polyethylene. Part I. Influence of Processing Conditions, and Part II. Effect of Polymer Grade," *Polymer Engineering & Science* **20**(9), June (1980).
5. S. Levy, "Extrusion Heads Change Shape to Produce Unusual Designs," *Plastics Machinery & Equipment* **8**(5), May (1979).
6. G. W. Shepherd, H. G. Clark, and G. W. Pearsall, "Extrusion of Polymer Tubing Using a Rotating Mandrel," *Polymer Engineering & Science* **16**(12), Dec. (1976).

7. W. J. Schrenk and T. Alfrey, "Some Physical Properties of Multilayer Films," *Polymer Engineering & Science* **9**(6), Nov. (1969).
8. T. Alfrey, E. F. Gurnee, and W. J. Schrenk, "Physical Optics of Iridescent Multilayered Plastic Films," *Polymer Engineering & Science* **9**(6), Nov. (1969).
9. S. Levy, "Improving the Performance of Extruded Products by Orientation," *Plastics Machinery & Equipment* **6**(12), Dec. (1977).
10. S. Levy, "Extrusion/Lamination Process Embeds Substrates in Thermoplastics," *Plastics Machinery & Equipment* **6**(3), March (1977).
11. S. Levy, "Processes and Equipment for In Line Post Extrusion Forming," *Plastics Machinery & Equipment,* **7**(8), Aug. (1978).

Glossary

adaptive control. A control system which changes the settings in response to changes in machine performance to bring the product back into specification. The shift is maintained so that the control has adapted to changing conditions.

adiabatic extruder operation. Operating condition in extruder where no heat passes in or out of the barrel wall.

Autoflex sheet die. Die in which the lip opening is controlled by thermal bolts responding to thickness sensors.

automatic control. Machine and line operating mode in which the machine output and line equipment operate in a feedback mode to continuously control dimensions.

automatic screen changer. Unit designed to remove one screen pack from the melt stream path and replace it with another pack when the pressure drop reaches a preset value.

biaxial orientation. A process for improvement of sheet properties by stretching the material in two right angle directions.

blow up ratio. The ratio of the diameter of the bubble to the die diameters in blown film extrusion.

blown film extrusion. A process for making thin plastics film by extruding a tube and blowing it up to many times the extruded diameter.

breaker plate. A disc with a number of holes drilled to permit the plastics material to flow. It is usually used to support screens. It is placed between the extruder barrel end and the die holder.

bushing, die. Tubing die section that forms the outside diameter of the tube.

capacitive sensor. Location or thickness sensor which uses the presence of the extrudate as a dielectric in a circuit to indicate thickness or proximity to extrudate.

capstan. Large drum device used to drive extrudates such as insulated wire by wrapping the extrudate around the drum to provide enough friction to get a nonslip drive.

check valve, foam extrusion. The valve used to control backflow of blowing agent in the gas injection method of foam extrusion.

choke plate. A single hole unit used between the end of the extruder barrel and the die holder to produce a controlled pressure drop in the plastics material flowing through the die.

choker bar, sheet die. Portion of sheet die used to selectively restrict flow to even out variations in sheet thickness.

closed loop dimension control system. A system which feeds back dimension information to adjust line speed to correct for dimension shifts.

computer control. A mode of machine operation where the extruder and line are under the control of a process computer which sets the parameters for the operation.

control loop. The signal circuit that provides feedback information for closed loop process control.

control system. The instruments and power controlling units which are used to hold machine temperature, pressure rate, and other parameters to the set values.

controllers. The discrete instruments used to control temperature, speed, and pressure in the machine operation.

cooling fixtures. Holding devices with air or water cooling for holding and setting extrusion shapes.

coextrusion. The extrusion process where two melt streams are combined in the die to make an extrusion of two plastics.

core pin. Tubing die center unit to form inner wall of the extrusion.

crammer feeder. A hopper unit which forces plastic material into the feed throat of the extruder.

crossflow. Flow in an extrusion die at right angles to the primary flow direction. Source of distortion in extrudate shape.

crosshead extrusion. An extrusion where the extrudate comes out of the machine at right angles to the barrel axis. It is used in wire covering and in some tube lines.

crystallinity. The amount of the structure of a polymer which is ordered as compared to the normal tangled amorphous state.

cut off units. Devices such as saws, shears, flying knives, and other devices for cutting extrusions to predetermined length.

dead band. Controller range in which the control is not affecting the temperature. The power is off and the temperature is allowed to drift.

die air vent. Passage in hollow pipe or profile die to permit the passage of air into interior of hollow extrusion.

die entry angle. Angle of convergence of melt entering the extrusion die lips.

die, extrusion. An orifice used to shape a plastics melt stream in the extrusion process.

die land, land length. The land is the straight section through an extrusion die. The land length is usually expressed as the ratio between the die opening and the length of the opening in the flow direction.

die spider. Legged unit to support the die section in the melt stream that forms the interior of a hollow section.

die swell. The enlargement of an extrudate over the dimensions of the die through which it is extruded.

downstream equipment. All of the auxiliary units used in an extrusion line after the die which are used to cool, shape, and control the extrudate.

drag flow. Mechanism by which material is caused to flow in a single-screw extruder and build pressure. The polymer must wet both screw and barrel to produce drag flow effects.

drawdown. Stretching the extrudate after it leaves the die by pulling faster than the rate at which it leaves the die.

dual durometer extrusion. Extruding a shape with a soft and a hard material, usually PVC materials.

elastic melt extruder. Extruder with twin discs which uses polymer melt elasticity effects to convey the plastics melt.

electric discharge machining (EDM). A method of shaping metal in die work by a process of spark erosion.

extrudate. The shaped material exiting from the extrusion die.

extruded foam. Extrusion process which expands the plastic mass by the formation of gas-filled cells.

extruder barrel. Main section of extruder in which the extruder screw turns.

extruder feedback control system. A method of controlling extruder output by adjustment of speed or back pressure to maintain a constant output rate.

extruder screw. The screw units used to propel plastics through a plastics extruder and to generate pressure on the melt.

extrusion. Process for making a product by forcing material through a die orifice.

extrusion valve. An adjustable restriction to melt flow which is used to control back pressure in an extruder.

feed chute. Entry hole in extruder barrel for plastics material.

finite element analysis. A mathematical procedure used to find flow effects as well as stress effects by dividing the analyzed space into discrete "finite" elements which are related to each other by constitutional equations. The method uses computer techniques to analyze stresses and flows in the absence of exact solutions.

free extrusion. Extrusion of tube or shape into cooling unit without the use of forming or controlling fixtures or sizers.

freeze line. The location on a blown film bubble where the polyolefin materials start to crystallize as a result of cooling of the bubble.

frictional heating. Heat generated by viscous mixing effects in the extruder.

gear pump extruder. Extruder using intermeshing gears to pump plastics.

Godet unit. A multiple roll drive unit usually used on monofilaments and strip where the extrudate is driven by the friction between the extrudate and the rolls. It is commonly used to orient monofilament.

hopper. Material holding unit attached to the extruder at the feed chute region to hold and feed plastics material into the extruder.

internal cooling mandrel. Tube sizing system using crosshead extrusion with cooled mandrel to size the inside diameter of the tubing.

mandrel. A term used to describe a sizing element either in a die or in a sizer which controls one dimension of the extrudate, usually the inside diameter.

melt fracture. The distortion of the surface of an extrudate after leaving a die orifice. The effect can range from minor ripples to severe distortion.

melt pressure. The pressure on the plastics material as it leaves the screw and enters the die.

melt strength. The characteristic of a plastics melt which consists in the ability to hold die shape and not sag after the material exits from the die.

melt temperature. The temperature of the plastics material in the extruder.

metering zone. Section of single-screw extruder which has a uniform flight depth that controls the rate of flow through the extruder into the die.

neckdown. Reduction of extrudate width caused by drawing down of sheet during extrusion.

non-Newtonian flow. Flow of a polymer material characterized by nonproportionality between shear rate and shear stress.

open loop control. Control of operation of an extruder or downstream unit where the speed or another parameter is set by the operator and is not adjusted by feedback information.

orientation. An extrusion process which increases the strength and stiffness of plastics by stretching or rolling the extrudate.

pinch draw rollers. A set of rollers used as pullers in the extrusion process. It is one form of puller unit.

plastication. The process of plasticizing a material.

plasticized. Refers to the condition of a plastics material when it has been softened for use in a melt processing operation such as extrusion.

plug flow. A flow condition through an extrusion die where the velocity across the melt stream is essentially constant.

profile extrusion dies. Dies which produce complex cross sections by extrusion exemplified by channels, gaskets, etc.

puller. Device used to remove the extrudate from the die region at a controlled rate. Usual form is either a pair of rollers or two belts in opposition with the extrudate pinched between.

ram extruder. Extrusion device using direct plunger pressure on the material for extrusion.

rheology, rheological characteristics. The study of the response of materials to stresses. The characteristics of materials in response to stresses.

rheometer. Device used to measure the melt properties of plastics materials. Both parallel plate and screw type units are used.

SCR (silicon-controlled rectifier) drive. A motor drive system that controls the speed of a DC motor by use of rectified pulses of power. The AC is converted to DC pulses of varying width under the effect of a control voltage generated by either a tachometer or a speed control. It is a method of regulating motor speed.

screen pack. Collection of screens in the breaker plate that controls pressure drop to the final die section and also filters debris from the extrusion stream.

shear heat generation. The heating effect caused by working the polymer in an extruder.

shear rate. The rate at which a material is undergoing deformation in response to a shear stress $V/H = dv/dh$.

shear stress. The force per unit area applied to a planar element inducing shear movement of a material.

sheet die restrictor bar. Unit used to even out flow along a sheet die to permit uniform thickness of extruded sheet.

sheet extrusion die. Die used to make plastics sheet of various widths.

single-screw extruder. Extruder using a single conveying screw to pump plastics.

static mixer blender. In-line unit that mixes the melt stream to make it uniform in temperature and composition by cutting and recombining flows of material without moving elements.

thixotrophy. A characteristic of material undergoing flow deformation where the viscosity increases drastically when the force inducing the flow is removed.

thread up. The procedure of starting an extrusion through the downstream equipment by threading it through the various downstream units.

three-roll stack. Interacting set of three metal cooling rollers used to cool and size sheet plastic in a sheet extrusion line.

tubing (pipe) die. A die with an outer bushing and inner pin with means of introducing the plastic melt in between to form round hollow product, i.e., tubing or pipe.

twin-screw extruder. Extruder using two intermeshing screws to convey plastics.

vacuum sizing. A shape-holding means used to set the dimensions of an extrudate by holding the material against a mandrel by means of a vacuum.

venturi cooling ring. A unit used to cool film and other extrusions by using a primary air stream to draw in additional room air by the venturi effect and improve cooling. It is also used to stabilize the extrudate while cooling.

viscosity, apparent viscosity. The ratio between the shear stress and the shear rate. A constant for a Newtonian fluid but variable for polymers. Apparent viscosity is the value over a narrow range for non-Newtonian fluids.

weir. Controlled flow aperture on the end of a water tank used to control the water overflow into the drain pans. The weirs are usually shaped to the extrudate.

winders. Units used to collect extrusions on spools or reels.

wire covering. A crosshead extrusion operation for the application of a plastics coating on wire.

Index